TOPICS IN OPERATOR THEORY

MATHEMATICAL SURVEYS • Number 13

TOPICS IN OPERATOR THEORY

EDITED BY

C. PEARCY

1974

AMERICAN MATHEMATICAL SOCIETY
PROVIDENCE, RHODE ISLAND

The articles in this volume were originally commissioned for the MAA STUDIES series, but were transferred by mutual agreement of the Association and the Society to MATHEMATICAL SURVEYS.

Library of Congress Cataloging in Publication Data **CIP**

Pearcy, C. 1935–

Topics in operator theory.

(Mathematical surveys, no. 13)

Includes bibliographies.

1. Linear operators. I. Title. II. Series: American Mathematical Society. Mathematical surveys, no. 13.

QA329.2.P4 515'.72 74-8254

ISBN 0-8218-1513-X

AMS (MOS) subject classifications (1970). Primary 47A02; Secondary 47B02.

TABLE OF CONTENTS

PREFACE

The articles in this volume are concerned with various aspects of the theory of bounded linear operators on Hilbert space. This area of mathematical research is presently experiencing a period of intense excitement, due, no doubt, to the fact that during the past year several remarkable advances have been made on hard problems in the field. One particular problem on which considerable progress has been made is the "invariant subspace problem." This is the question whether every (bounded, linear) operator T on a separable, infinite-dimensional, complex Hilbert space $\mathcal{H}$ maps some (closed) subspace different from (0) and $\mathcal{H}$ into itself. It is therefore highly appropriate that the first and last of the five expository articles in this volume deal with invariant subspaces.

The main theme of the first article, by Donald Sarason, may be summarized as follows. If T is a bounded linear operator on $\mathcal{H}$, then the collection of all subspaces $\mathcal{M}$ of $\mathcal{H}$ such that T maps $\mathcal{M}$ into itself forms a complete lattice, denoted by $\operatorname{Lat}(T)$, under the inclusion ordering. It is presently impossible to determine, in general, the lattice-theoretic structure of $\operatorname{Lat}(T)$. (This is not surprising, since we cannot even say whether the possibility $\operatorname{Lat}(T) = \{(0), \mathcal{H}\}$ is realizable.) There are a few operators T, however, for which the structure of $\operatorname{Lat}(T)$ is known, and Sarason's article is devoted to a discussion of such operators and their associated invariant subspace lattices. The presentation is extremely lucid, and one of the interesting features of the exposition is the interplay between operator theory and classical analysis that is so often found in Sarason's work.

The second article in this volume is concerned with weighted shift operators. Interest focuses on this class of operators because it is a particularly simple class to define. Let $\{e_n\}_{n=0}^{\infty}$ be an orthonormal basis for $\mathcal{H}$, and let $\alpha = \{\alpha_n\}_{n=0}^{\infty}$ be a bounded sequence of complex numbers (the weight sequence). It is easy to see that there exists a unique bounded operator T_α on $\mathcal{H}$ such that $T_\alpha e_n = \alpha_n e_{n+1}$ for every nonnegative integer n, and that the adjoint operator T_α^* is determined by the equations $T_\alpha^* e_0 = 0$ and $T_\alpha^* e_n = \overline{\alpha_{n-1}} e_{n-1}$, $n > 0$. The operators T_α and T_α^* are typical examples of *forward* and *backward weighted* *unilateral shifts,* respectively. A different type of weighted shift is obtained by considering an orthonormal bases $\{f_n\}_{n=-\infty}^{\infty}$ for H indexed by the set of all integers, and an associated (bounded) weight sequence $\beta = \{\beta_n\}_{n=-\infty}^{\infty}$. In this case, the unique bounded operator T_β on H that satisfies the equation $T_\beta f_n = \beta_n f_{n+1}$ for all integers n is a typical *forward weighted bilateral shift,* and a typical backward weighted bilateral shift can be defined analogously. Allen Shields has taken essentially all of the information presently known about weighted shift operators (with scalar weights) and incorporated it into his comprehensive article. A central theme of the exposition is the interplay between weighted shift operators and analytic function theory, and, as an added bonus for the reader, the article contains a list of thirty-two interesting research problems.

The third article in this volume is an exposition by Arlen Brown of the theory of spectral multiplicity for normal operators on Hilbert space. The problem treated arises as follows. In general, one wants to know when two operators T_1 and T_2 acting on Hilbert spaces K_1 and K_2, respectively, are unitarily equivalent (i. e., when there is an inner product preserving isomorphism φ of K_1 onto K_2 such that $\varphi T_1 \varphi^{-1} = T_2$). Unitary equivalence is the analog for operators of the concept of isomorphism for groups, rings, etc. The problem is usually attacked by trying to attach to each operator T in a fixed class of operators an indexed family of objects $\{O_\lambda(T)\}_{\lambda \in \Lambda}$ with the property that two operators T_1 and T_2 in the class are unitarily equivalent if and only if $O_\lambda(T_1) = O_\lambda(T_2)$ for all indices λ in Λ. Such a collection $\{O_\lambda(T)\}_{\lambda \in \Lambda}$ is called a *complete set of unitary invariants* for the class of operators. There are a few classes of operators for which a "reasonable" complete set of unitary invariants is known; in this connection see the bibliography of Brown's article. The theory of spectral multiplicity furnishes a reasonable complete set of unitary invariants for normal operators, and this set of invariants can be obtained in several different ways. One may consider the (commutative) C^*-algebra or von Neumann algebra generated by a given normal operator, and study the unitary equivalence problem for such algebras. In this approach, the solution of the unitary equivalence problem for normal operators becomes a corollary of the solution of the problem for commutative C^*-algebras or von Neumann algebras. One may also proceed more directly and focus attention on the concept of a spectral measure. It is this last approach that Brown follows, and his clear presentation of this circle of ideas should lead to a better understanding of multiplicity theory by beginners and experts alike.

The fourth article in this volume, by R. G. Douglas, is concerned with canonical models for operators. The central underlying idea here is that given any contraction operator T on H (i. e., any operator T satisfying $\|T\| \leqslant 1$), there is a canonical construction that associates with T an operator M_T that is unitarily equivalent to T, called its "canonical model." Thus, one may study T by studying M_T instead, and this is one of the themes of the book *Harmonic analysis of operators on Hilbert space,* by Sz.-Nagy and Foiaș. Douglas, who has contributed significantly to the geometrization of the theory of canonical models, exposes in his article various important components of this theory, and thereby gives the reader much insight into its successes and failures.

The final article in this volume, written by Allen Shields and myself, is a survey of some invariant-subspace theorems that resulted from the brilliant and elegant method of proof introduced by Victor Lomonosov early in 1973. It would indeed be surprising if further study and refinement of this technique does not lead to additional progress on the invariant subspace problem.

CARL PEARCY

ANN ARBOR, MICHIGAN
DECEMBER 1973

Mathematical Surveys
Volume 13
1974

I

INVARIANT SUBSPACES

BY

DONALD SARASON

AMS (MOS) subject classifications (1970). Primary 47A15; Secondary 30A78.
Preparation of this paper supported in part by a Sloan Foundation Fellowship.

DONALD SARASON

1. Introduction. This article surveys certain aspects of the theory of invariant subspaces of operators. We shall limit our attention to bounded operators on Hilbert spaces, and by a subspace we shall always mean a closed subspace. A subspace is called invariant under an operator if the operator maps the subspace into itself.

The investigation of invariant subspaces is a natural first step in the attempt to understand the structure of operators. The powerful structure theorems that are known for finite-dimensional operators (the Jordan form) and normal operators (the spectral theorem) provide, in essence, decompositions into invariant subspaces of special kinds. No comparable theorem exists, or is at present imaginable, for the general operator on an infinite-dimensional Hilbert space. Indeed, it is still an open question whether every operator on an infinite-dimensional Hilbert space has an invariant subspace other than the zero subspace and the whole space.

The preceding question, which is known as the "invariant subspace problem," provides much of the motivation for current interest in invariant subspaces. Although the problem has attracted the attention of many powerful mathematicians, and although many partial results are known, a complete solution at present appears remote. Most experts expect an eventual negative answer; that is, they suspect the existence of a Hilbert space operator with no invariant subspaces other than the two trivial ones. (However, I have never heard anyone put forth a convincing reason for this expectation.)

There is another motivation for current interest in invariant subspaces. Although the general operator remains a mystery, one can say quite a bit about the invariant subspaces of a handful of specific operators, and results of this nature are often connected with interesting theorems in analysis. This was first realized approximately twenty years ago by Beurling, who gave a complete classification of the invariant subspaces of the unilateral shift operator (the operator on l^2 defined by $(c_0, c_1, c_2, \cdots) \longrightarrow (0, c_0, c_1, \cdots)$).

My emphasis in this article will be on bringing out the connections with analysis alluded to in the preceding paragraph, and I shall say almost nothing about the existence problem. The reader can find information on the latter problem in the papers **[1]**, **[3]**, **[15]**, **[40]**, **[41]** and in the book **[17]**. The contents of the article are previewed in greater detail in the next section.

The article should be understandable to anyone who has taken basic courses in measure theory, functional analysis, and functions of a complex variable. Given these prerequisites, the exposition is nearly self-contained as far as proofs are concerned; the few exceptions to this should not cause serious difficulties. On the other hand, I have throughout the article left straightforward verifications to the reader. A student who is encountering some of the ideas presented here for the first time would probably benefit from supplying omitted details.

The bibliography at the end of the article contains only papers and books referred to in the text; I have made no attempt to compile an exhaustive list of references. Still, the bibliography will provide an ample starting point for anyone who wishes to pursue the subject further. The present article was originally written in 1970 but its publication has been unexpectedly delayed. In the meantime, there has been considerable progress in the understanding of invariant subspaces. For an up-to-date exposition and a substantial bibliography, I recommend the recent monograph by H. Radjavi and P. Rosenthal entitled *Invariant subspaces* (Ergebnisse der Mathematik und ihrer Grenzgebiete, Band 77, Springer-Verlag, Berlin and New York, 1973).

The ideas I have used in writing this article come from many sources and it would be impossible for me to trace them all back to their beginnings. The general point of view I have adopted has been greatly influenced by Paul Halmos, my teacher, and Henry Helson, my colleague. I should like to express my gratitude to these gentlemen.

2. Some immediate observations. Let A be an operator on a Hilbert space H. We allow H to be finite dimensional but, to avoid trivialities, we shall exclude the case where H is one dimensional. The reader will easily verify that the invariant subspaces of A^* are the orthogonal complements of the invariant subspaces of A. A subspace S is said to *reduce* A if it is invariant under both A and A^*, in other words, if both S and $S^\perp$ are invariant under A. When this happens A can be written as the direct sum of its restriction to S and its restriction to $S^\perp$. It is easy to see that S reduces A if and only if the orthogonal projection onto S commutes with A.

An operator may have no reducing subspaces other than the zero subspace and the whole space; the operator is then said to be *irreducible*. The unilateral shift operator mentioned in §1 is irreducible, as are the analogous operators on finite-dimensional Hilbert spaces. A normal operator, on the other hand, is always reducible. This follows from the spectral theorem, which assures us of the existence of many orthogonal projections that commute with the operator. In the next section we shall study reducing subspaces of normal operators in some detail.

If S_1 and S_2 are invariant subspaces of A, then the span of S_1 and S_2 and the intersection of S_1 and S_2 are also invariant under A. Hence the invariant subspaces of A form a sublattice of the lattice of all subspaces of H. We denote the family of all invariant subspaces of A by Lat A. The lattice Lat A is complete; that is to say, every collection of invariant subspaces of A has a least upper bound (their span) and a greatest lower bound (their intersection).

The preceding observations prompt the question: Which complete lattices are isomorphic to Lat A for some A? Such lattices are said to be *attainable*. A special case of the above question asks whether the trivial lattice $\{0, 1\}$ is attainable. That, of course, is the invariant subspace problem.

A few examples of attainable lattices are the following:

(1) the lattice formed by the integers between 0 and n ($n > 1$), with their natural order (attained by a cyclic nilpotent operator on an n-dimensional Hilbert space);

(2) the nonpositive integers with $-\infty$ appended (attained by certain unilateral weighted shifts; see the article by A. L. Shields in this volume);

(3) the nonnegative integers with $+\infty$ appended (attained by the adjoints of the weighted shifts referred to above);

(4) the closed unit interval (attained by the Volterra operator; see §8);

(5) the lattice of closed subsets of the closed unit interval (see §9);

(6) the lattice of Lebesgue measurable subsets of the unit interval, modulo null sets (see §3).

It is unknown, for example, whether the integers with $+\infty$ and $-\infty$ appended form an attainable lattice. (Further results and problems on attainability, along with references, can be found in **[34]**.)

Along with invariant subspaces of operators it is natural to consider invariant subspaces of families of operators, and especially of algebras of operators. For F a family of operators on H, we let Lat F denote the family of subspaces of H that are invariant under every operator in F. Obviously, the invariant subspaces of F are invariant under sums and products of operators in F and hence under the operator algebra generated by F. Also, because a subspace is weakly closed, an invariant subspace of F is invariant under every operator that lies in the closure of F with respect to the weak operator topology. The last two observations combine to give a simple but often useful proposition.

PROPOSITION 1. *Let* F *be a family of operators on* H *and let* A *be the weak closure of the algebra generated by* F *and the identity operator. Then* $\operatorname{Lat} \mathsf{A} = \operatorname{Lat} \mathsf{F}$.

The problem of finding converses to this proposition has recently attracted attention. The question can be posed in the following way: *Let* A *be a weakly closed algebra of operators on* H, *and let* B *be an operator on* H *such that* $\operatorname{Lat} \mathsf{A} \subset \operatorname{Lat} B$. *Under what additional conditions can one conclude that* B *belongs to* A? Simple examples show that some additional condition is needed, even when $\dim H = 2$. We shall look into this question in §4.

At this point we have touched on most of the topics that will concern us. The remaining ones relate to the theorem of Beurling mentioned in §1. We shall prove this theorem in §6 after an investigation in §5 of the invariant subspaces of unitary operators. The latter investigation will bring us into contact with a number of interesting results in analysis.

3. Reducing subspaces of normal operators. If A is a Hilbert space operator, then the reducing subspaces of A are the ranges of the orthogonal projections P such that $AP = PA$. If one takes the adjoint of both sides of the last equality one obtains the equality $PA^* = A^*P$. Thus, the problem of finding the reducing subspaces of A is contained in the problem of finding all operators that commute with both A and A^*.

In the following theorem, the preceding problem is solved for an especially simple class of operators.

THEOREM 1. *Let* m *be a finite, positive, compactly supported Borel measure in the complex plane, and let* A *be the operator on* $L^2(m)$ *of multiplication by* z $((Af)(z) = zf(z))$. *Then the operators that commute with* A *and* A^* *are precisely the operators on* $L^2(m)$ *of multiplication by the functions in* $L^\infty(m)$.

The operator A of Theorem 1 is, to within unitary equivalence, the most general cyclic normal operator. This is easily proved from the spectral theorem; see for example [**10**, p. 910].

One half of Theorem 1 is trivial: If A is as described and φ is in $L^\infty(m)$, then multiplication by φ obviously defines an operator on $L^2(m)$ that commutes with A and A^*. The other half of the theorem, although less obvious, is still easy to prove.

Let B be an operator that commutes with A and A^*. Let $\varphi = B1$, where 1 denotes the function in $L^2(m)$ that is constantly equal to 1. We shall show that φ is essentially bounded and that B is the operator on $L^2(m)$ of multiplication by φ.

Consider first a polynomial $p = p(z, \bar{z})$ in the two variables z and $\bar{z}$. Since B commutes with A and A^* it commutes with $p(A, A^*)$. Hence

$$(1) \qquad Bp = Bp(A, A^*)1 = p(A, A^*)B1 = p(A, A^*)\varphi = \varphi p.$$

Now consider any f in $L^2(m)$. By the Stone-Weierstrass theorem, the polynomials in z and $\bar{z}$ are uniformly dense in the space of complex-valued continuous functions on the support of m, and hence they are dense in $L^2(m)$. Therefore there is a sequence $\{p_n\}$ of polynomials in z and $\bar{z}$ such that $p_n \longrightarrow f$ both in the norm of $L^2(m)$ and almost everywhere with respect to m. By (1) we have $Bp_n = \varphi p_n$ for every n. As $n \longrightarrow \infty$ we have $Bp_n \longrightarrow Bf$ in $L^2(m)$ (by the continuity of B) and $\varphi p_n \longrightarrow \varphi f$ almost everywhere. Hence $Bf = \varphi f$.

It only remains to show that φ is essentially bounded. Let ϵ be a positive number, and let E_ϵ be the set where $|\varphi|^2 \geqslant \|B\|^2 + \epsilon$. Let χ_ϵ be the characteristic function of E_ϵ. Then χ_ϵ belongs to $L^2(m)$ and has norm $m(E_\epsilon)^{1/2}$. Hence $\|B\chi_\epsilon\|^2 \leqslant \|B\|^2 \|\chi_\epsilon\|^2 = \|B\|^2 m(E_\epsilon)$. But also

$$\|B\chi_\epsilon\|^2 = \|\varphi\chi_\epsilon\|^2 = \int_{E_\epsilon} |\varphi|^2 \, dm \geqslant (\|B\|^2 + \epsilon)m(E_\epsilon).$$

Combining the last two inequalities, we obtain

$$(\|B\|^2 + \epsilon)m(E_\epsilon) \leqslant \|B\|^2 m(E_\epsilon),$$

which is only possible if $m(E_\epsilon) = 0$. As the latter holds for all $\epsilon > 0$ we may conclude that φ is essentially bounded, with $\|\varphi\|_\infty \leqslant \|B\|$. The proof of the theorem is complete. (We note that the inequality $\|B\| \leqslant \|\varphi\|_\infty$ is trivial, so actually $\|\varphi\|_\infty = \|B\|$.)

The operator B is a projection if and only if $\varphi^2 = \varphi$ almost everywhere, in other words, if and only if $\varphi = 0$ or 1 almost everywhere. The projections that commute with A are thus the multiplication operators induced by the characteristic functions in $L^\infty(m)$. This provides the following characterization of the reducing subspaces of A.

COROLLARY 1. *Let A be as in Theorem 1. Then the reducing subspaces of A are the subspaces $\chi_E L^2(m)$ with E a Borel subset of the plane.*

Here, χ_E denotes the characteristic function of E, and $\chi_E L^2(m)$ stands for the subspace of all functions $\chi_E f$ with f in $L^2(m)$.

The above corollary establishes a one-to-one correspondence between the family of reducing subspaces of A and the family of equivalence classes modulo m of Borel subsets of the plane. The latter family is commonly called the *measure algebra* of m. It is a Boolean algebra under the ordinary set theoretic operations, so in particular it is a lattice. The correspondence in question is a lattice isomorphism.

If the measure m is supported by the real line then the operator A is selfadjoint, so that every invariant subspace of A is a reducing subspace. For this case, therefore, Corollary 1 provides a complete description of Lat A. In particular, we have

COROLLARY 2. *If m is a finite, positive, compactly supported Borel measure on the real line, then the measure algebra of m is an attainable lattice.*

The class of measure algebras covered by Corollary 2 is more widely representative than one might at first suspect; see [**13**, §41].

There is a generalization of Theorem 1 which leads to a classification of the reducing subspaces of the most general normal operator on a separable Hilbert space. I shall describe this generalization but shall not give its proof. We must employ some simple vector-valued function theory, a thorough discussion of which can be found in [**20**, Chapter III].

Let m be as before, and let $\mathcal{V}$ be a separable Hilbert space (finite or infinite dimensional). We consider $\mathcal{V}$-valued functions on the complex plane, that is, functions F from the complex plane into $\mathcal{V}$. Such an F is said to be *Borel measurable* if the scalar function $(F(z), x)$ is Borel measurable for each x in $\mathcal{V}$. When this happens, the scalar function $\|F(z)\|$ is also Borel measurable, so $\int \|F(z)\|^2\, dm(z)$ is well defined. The space of functions F for which the latter integral is finite is denoted by $L^2(m; \mathcal{V})$. It is a separable Hilbert space under the ordinary pointwise algebraic operations and the inner product $(F, G) = \int (F(z), G(z))\, dm(z)$, provided one identifies functions that are equal almost everywhere modulo m.

Let the operator A on $L^2(m; \mathcal{V})$ be defined by $(AF)(z) = zF(z)$. This operator is a direct sum of copies of the operator of Theorem 1, the number of summands being the dimension of $\mathcal{V}$. We ask: What are the operators that commute with A and A^*?

It is easy to pick out some operators that commute with A and A^*. The multiplication operators induced in the obvious way by the functions in $L^\infty(m)$ clearly have this property. If $\mathcal{V}$ is not one dimensional then there are others.

For example, if T is any operator on $\mathcal{V}$ then T induces an operator on $L^2(m; \mathcal{V})$, namely, the operator that sends $F(z)$ to $TF(z)$, and this operator commutes with A and A^*. There are still more, because the operators that commute with A and A^* form an algebra. If we take the algebra generated by the operators mentioned so far, we obtain the multiplication operators on $L^2(m; \mathcal{V})$ induced by certain operator-valued functions. In fact, it is clear that any operator on $L^2(m; \mathcal{V})$ that is induced (in the obvious way) by an operator-valued function will commute with A and A^*. The operator-valued functions that induce multiplication operators on $L^2(m; \mathcal{V})$ form an operator-valued analogue of the space $L^\infty(m)$ which I now define.

Let $\mathcal{B}$ denote the space of operators on $\mathcal{V}$. A $\mathcal{B}$-valued function Φ on the complex plane is said to be *Borel measurable* if the scalar function $(\Phi(z)x, y)$ is Borel measurable for each pair of vectors x, y in $\mathcal{V}$. When this happens the scalar function $\|\Phi(z)\|$ is also Borel measurable. The space of functions Φ for which the latter function is essentially bounded is denoted by $L^\infty(m; \mathcal{B})$. This space is a Banach algebra under the pointwise algebraic operations and the essential supremum norm (provided functions are identified that are equal almost everywhere modulo m). If Φ is in $L^\infty(m; \mathcal{B})$ and F is in $L^2(m; \mathcal{V})$, then the $\mathcal{V}$-valued function $\Phi(z)F(z)$ is Borel measurable. Clearly, therefore, the latter function belongs to $L^2(m; \mathcal{V})$, and its norm does not exceed the norm of Φ times the norm of F. Thus, each function in $L^\infty(m; \mathcal{B})$ induces a multiplication operator on $L^2(m; \mathcal{V})$. The generalization of Theorem 1 reads as follows:

THEOREM 1$'$. *The operators on $L^2(m; \mathcal{V})$ that commute with A and A^* are precisely the multiplication operators induced by the functions in $L^\infty(m; \mathcal{B})$.*

Theorem 1 is the special case where $\mathcal{V}$ is one dimensional. The reader should have no difficulty in proving Theorem 1$'$ from Theorem 1 in the case where $\mathcal{V}$ is finite dimensional. In the case where $\mathcal{V}$ is infinite dimensional, technical difficulties arise. Nevertheless, the above proof of Theorem 1 can be adapted to handle the general case; see [**26**, §4] for the details.

If Φ is in $L^\infty(m; \mathcal{B})$, then the operator on $L^2(m; \mathcal{V})$ induced by Φ is a projection if and only if $\Phi(z)$ is a projection for almost every z modulo m. Hence the reducing subspaces of A are in one-to-one correspondence with the projection valued functions in $L^\infty(m; \mathcal{B})$ (where two functions that are equal almost everywhere are identified).

We are only one step away from a description of the reducing subspaces of the most general normal operator on a separable Hilbert space. We assume now that V is infinite dimensional, and we select an increasing sequence $V_1, V_2, V_3, \cdots$ of subspaces of V such that V_n has dimension n and $\bigcup V_n$ is dense in V. For convenience we let $V_\infty = V$. Let ν be a Borel measurable function on the plane whose values are positive integers or ∞. Let $L^2(m; V, \nu)$ denote the space of functions F in $L^2(m; V)$ such that $F(z)$ belongs to $V_{\nu(z)}$ for all z. Clearly, $L^2(m; V, \nu)$ is a reducing subspace of the operator A; we let N denote the restriction of A to this subspace. The operator N is, to within unitary equivalence, the most general normal operator on a separable Hilbert space. This result, which can be viewed as a sharpening of the spectral theorem, is proved, essentially, in [**10**, Chapter X, §5]. See also the article by A. Brown in this volume.

Let $L^\infty(m; \mathcal{B}, \nu)$ be the space of functions Φ in $L^\infty(m; \mathcal{B})$ such that, for each z, the ranges of $\Phi(z)$ and $\Phi(z)^*$ are contained in $V_{\nu(z)}$. The functions in $L^\infty(m; \mathcal{B}, \nu)$ induce multiplication operators on $L^2(m; V, \nu)$ and, by Theorem 1′, these are precisely the operators that commute with N and N^*. Hence the reducing subspaces of N are in one-to-one correspondence with the projection valued functions in $L^\infty(m; \mathcal{B}, \nu)$ (functions agreeing almost everywhere being identified).

As before, if m is supported by the real line, then N is selfadjoint, so we obtain a complete description of Lat N.

It should be added that, by a theorem of Fuglede [**33**], if a normal operator N commutes with an operator B, then N^* also commutes with B. The above remarks thus constitute a description of the commutant of the most general normal operator on a separable Hilbert space.

4. Invariant subspaces and operator algebras. Let H be a Hilbert space and let $\mathcal{B}(H)$ denote the algebra of all operators on H. By a *subalgebra* of $\mathcal{B}(H)$ we shall always mean a subalgebra that contains the identity operator. A subalgebra $\mathcal{A}$ of $\mathcal{B}(H)$ is said to be ***reflexive*** if, for B in $\mathcal{B}(H)$, the inclusion Lat $\mathcal{A} \subset$ Lat B implies that B belongs to $\mathcal{A}$. By Proposition 1 (§2), all reflexive algebras are weakly closed.

The oldest theorem about reflexive algebras is the double commutant theorem. We recall that if $\mathcal{A}$ is a subalgebra of $\mathcal{B}(H)$, then the commutant of $\mathcal{A}$, denoted $\mathcal{A}'$, is the family of operators that commute with every operator in $\mathcal{A}$, and the double commutant of $\mathcal{A}$, denoted $\mathcal{A}''$, is the commutant of $\mathcal{A}'$. The double commutant theorem can be stated, in standard fashion, as follows: *If* $\mathcal{A}$ *is a weakly closed selfadjoint subalgebra of* $\mathcal{B}(H)$, *then* $\mathcal{A}'' = \mathcal{A}$. A proof of this theorem can be found in [**8**, Chapter I, §3.4].

To see the connection between the double commutant theorem and reflexive algebras, let $\mathcal{A}$ be any selfadjoint subalgebra of $\mathcal{B}(H)$. Then $\mathcal{A}$ is reduced by all of its invariant subspaces, so the invariant subspaces of $\mathcal{A}$ are the ranges of the projections in $\mathcal{A}'$. Hence if B is in $\mathcal{A}''$ then $\mathrm{Lat}\,\mathcal{A} \subset \mathrm{Lat}\,B$. Suppose, conversely, that B is an operator in $\mathcal{B}(H)$ such that $\mathrm{Lat}\,\mathcal{A} \subset \mathrm{Lat}\,B$. Then B commutes with all projections in $\mathcal{A}'$. Now $\mathcal{A}'$ is a weakly closed algebra, and therefore it contains the spectral projections of each of its selfadjoint operators [8, p. 3]. But for B to commute with a selfadjoint operator it is sufficient that B commute with the spectral projections of that operator. Hence B commutes with every selfadjoint operator in $\mathcal{A}'$. Since $\mathcal{A}'$ is selfadjoint, this means that B belongs to $\mathcal{A}''$. We see therefore that the operators B in $\mathcal{A}''$ are precisely those satisfying $\mathrm{Lat}\,\mathcal{A} \subset \mathrm{Lat}\,B$. Accordingly, the double commutant theorem can be restated as follows: *Every weakly closed selfadjoint subalgebra of* $\mathcal{B}(H)$ *is reflexive.*

Thus, the double commutant theorem closes the subject of selfadjoint reflexive algebras (in much the same way as the Stone-Weierstrass theorem closes the subject of uniform approximation by selfadjoint algebras of continuous complex-valued functions on compact Hausdorff spaces). We turn to the subject of nonselfadjoint reflexive algebras. The following theorem is a sample.

THEOREM 2. *Let* m *be a finite, positive, compactly supported Borel measure on the complex plane, and let* $\mathcal{A}_0$ *be a subalgebra of* $L^\infty(m)$ *which contains the constant functions and is closed in the weak-star topology of* $L^\infty(m)$. *Let* $\mathcal{A}$ *be the algebra of multiplication operators on* $L^2(m)$ *induced by the functions in* $\mathcal{A}_0$. *Then* $\mathcal{A}$ *is reflexive.*

We recall that $L^\infty(m)$ is the dual of the space $L^1(m)$; the weak-star topology of $L^\infty(m)$ is the topology that arises from this duality. Corresponding to each finite set $\{f_1, \cdots, f_q\}$ of nonzero vectors in $L^1(m)$ we have a weak-star neighborhood of the origin in $L^\infty(m)$ defined by

$$V(f_1, \cdots, f_q) = \left\{\psi \in L^\infty(m): \left|\int \psi f_k \, dm\right| < 1, k = 1, \cdots, q\right\}.$$

The family of all such sets $V(f_1, \cdots, f_q)$ is a weak-star neighborhood base for the origin.

To prove Theorem 2, let B be an operator on $L^2(m)$ such that $\mathrm{Lat}\,\mathcal{A} \subset \mathrm{Lat}\,B$. Then, in particular, B leaves invariant the subspace $\chi_E L^2(m)$ for any Borel set E. Hence B commutes with the operators on $L^2(m)$ of multiplication by the functions χ_E. This implies that B commutes with all the multiplication operators on $L^2(m)$ induced by the functions in $L^\infty(m)$. Therefore, by Theorem 1, B must be the operator on $L^2(m)$ of multiplication by some function φ in $L^\infty(m)$.

It remains to show that φ belongs to A_0. Since A_0 by assumption is weak-star closed, it will be enough to show that every weak-star neighborhood of φ meets A_0. If $V(f_1, \cdots, f_q)$ is as above, then $\varphi + V(f_1, \cdots, f_q)$ is a typical basic weak-star neighborhood of φ. Thus we need only show that there is a ψ in A_0 such that $\psi - \varphi$ is in $V(f_1, \cdots, f_q)$. Let $f = |f_1| + \cdots + |f_q|$ and $g = f^{1/2}$. Then g is in $L^2(m)$, and the $L^2(m)$-closure of $\mathsf{A}_0 g$ is an invariant subspace of A. This subspace is therefore invariant under B, so it contains $Bg = \varphi g$. Hence there is a ψ in A_0 such that $\|\psi g - \varphi g\| < 1/\|g\|$. For any $k = 1, \cdots, q$ we have

$$\left|\int (\psi - \varphi) f_k \, dm\right| \leqslant \int |\psi - \varphi| f \, dm = \int (\psi - \varphi) gh \, dm,$$

where

$$\begin{aligned} h &= g|\psi - \varphi|/(\psi - \varphi) \quad \text{where } \psi - \varphi \neq 0, \\ &= 0 \qquad\qquad\qquad\quad\;\; \text{where } \psi - \varphi = 0. \end{aligned}$$

By Schwarz's inequality,

$$\int (\psi - \varphi) gh \, dm \leqslant \|\psi g - \varphi g\| \; \|h\| < (1/\|g\|) \cdot \|h\| \leqslant 1,$$

and so $\psi - \varphi$ is in $V(f_1, \cdots, f_q)$, as desired. The proof of Theorem 2 is complete.

We remark, as the reader probably already has, that if $L^\infty(m)$ is identified with the algebra of multiplication operators it induces on $L^2(m)$, then the weak-star topology identifies with the weak operator topology. Accordingly, the assumption in Theorem 2 that A_0 is weak-star closed is equivalent to the assumption that A is weakly closed.

Theorem 2 is a special case of the following more general result:

THEOREM 2′. *Let H be a Hilbert space and A a commutative weakly closed subalgebra of $\mathsf{B}(H)$ consisting of normal operators. Then A is reflexive.*

I sketch the proof. Let B be an operator on H such that Lat $\mathsf{A} \subset$ Lat B.

Let $\mathcal{B}$ be the weakly closed selfadjoint operator algebra generated by $\mathcal{A}$. An application of the double commutant theorem shows that B belongs to $\mathcal{B}$. The algebra $\mathcal{B}$ is commutative by the theorem of Fuglede mentioned at the end of §3.

To show that B belongs to $\mathcal{A}$ it is enough to show that every weak neighborhood of B meets $\mathcal{A}$. The following lemma enables one to accomplish this by a slight modification of the reasoning in the proof of Theorem 2.

LEMMA. *Let $\mathcal{B}$ be a commutative selfadjoint algebra of operators on a Hilbert space H. Then, given finitely many vectors $g_1, \cdots, g_q, h_1, \cdots, h_q$ in H, there is a vector g in H such that $|(Ag_k, h_k)| \leqslant \|Ag\|$ for $k = 1, \cdots, q$ and all A in $\mathcal{B}$.*

The lemma also can be obtained by a modification of the reasoning in the proof of Theorem 2, because any such algebra $\mathcal{B}$ can be represented as an algebra of multiplication operators on the L^2-space of a suitable measure [**39**, §3].

Theorem 2′ is due to the author [**37**]. The above proof is due to R. Goodman [**12**], who independently discovered the following consequence of Theorem 2′.

COROLLARY. *For a normal operator A, the equality* $\operatorname{Lat} A = \operatorname{Lat} A^*$ *holds if and only if A^* lies in the weak closure of the set of polynomials in A.*

If $\mathcal{A}$ is a commuting algebra of normal operators then, by Fuglede's theorem, $\operatorname{Lat} \mathcal{A}$ contains the ranges of the spectral projections of all operators in $\mathcal{A}$. Thus, the algebras of Theorem 2′ are comparatively rich in invariant subspaces. Recently a number of theorems have been proved concerning the reflexivity of algebras which are poor in invariant subspaces. The following result of Radjavi and Rosenthal [**31**] is one of the nicest: *Let $\mathcal{A}$ be a weakly closed subalgebra of $\mathcal{B}(H)$ which contains a maximal commutative selfadjoint subalgebra of $\mathcal{B}(H)$. If* $\operatorname{Lat} \mathcal{A}$ *is linearly ordered, then $\mathcal{A}$ is reflexive.* This improves an earlier theorem of Arveson [**2**] which, in place of the assumption that $\operatorname{Lat} \mathcal{A}$ is linearly ordered, makes the assumption that $\operatorname{Lat} \mathcal{A}$ is trivial. (The conclusion of Arveson's theorem is thus that $\mathcal{A} = \mathcal{B}(H)$.) Theorems related to the one of Radjavi and Rosenthal have been obtained by Nordgren [**28**], by Nordgren, Radjavi and Rosenthal [**29**], [**30**], and by Davis, Radjavi and Rosenthal [**7**].

In concluding this section I wish to mention the following question: *If A and B are commuting Hilbert space operators such that* $\operatorname{Lat} A \subset \operatorname{Lat} B$, *must B belong to the weakly closed algebra generated by A?* Theorem 2′ provides an affirmative answer when A is normal; in that case (but not generally) the condition that A and B commute is a consequence of the inclusion $\mathrm{Lat}\, A \subset \mathrm{Lat}\, B$. It is intriguing that the question also has an affirmative answer in the finite-dimensional case.

THEOREM 3. *Let A and B be operators on a finite-dimensional Hilbert space H such that $AB = BA$ and $\mathrm{Lat}\, A \subset \mathrm{Lat}\, B$. Then B is a polynomial in A.*

To prove the theorem we write H as a (perhaps nonorthogonal) direct sum of subspaces $H_1, \cdots, H_n$ which are invariant under A and on which A acts cyclically. That this is possible follows, for example, from the Jordan decomposition. For $k = 1, \cdots, n$ let h_k be a cyclic vector for $A|H_k$, and let $h = h_1 + \cdots + h_n$. Let S be the invariant subspace of A generated by h. Then S is invariant under B, so there is a polynomial p such that $Bh = p(A)h$. Thus

$$Bh_1 + \cdots + Bh_n = p(A)h_1 + \cdots + p(A)h_n.$$

Since the subspaces $H_1, \cdots, H_n$ are linearly independent and invariant under B, it follows that $Bh_k = p(A)h_k$ for each k. Thus the operators $B|H_k$ and $p(A)|H_k$ agree on a cyclic vector for $A|H_k$. As these operators both commute with $A|H_k$ it follows that they are identical, and we may conclude that $B = p(A)$.

Theorem 3 is due to Brickman and Fillmore [**5**, Theorem 10].

5. Unitary operators. In this section we investigate the invariant subspaces of unitary operators. The results we shall obtain have many interesting implications, a few of which we shall explore in this and the following three sections. In particular, the present section contains a proof of the famous theorem of F. and M. Riesz on analytic measures.

The need to analyze the invariant subspaces of unitary operators arises, among other ways, in connection with the theory of unitary dilations. The latter theory is treated in the article by R. G. Douglas in this volume, and the reader will find further information on our subject there.

We begin with an observation concerning the invariant subspaces of arbitrary operators. An invariant subspace of an operator will be called *irreducible* if it contains no nontrivial reducing subspaces of the operator.

PROPOSITION 2. *Let A be a Hilbert space operator and S an invariant subspace of A. Then S has a unique direct sum decomposition $S = S_{\mathrm{II}} \oplus S_{\mathrm{I}}$,* *where* $S_{\rm II}$ *is a reducing subspace of* A *and* $S_{\rm I}$ *is an irreducible invariant subspace of* A.

The proposition is all but trivial. We define $S_{\rm II}$ to be the closed linear span of all the subspaces of S that reduce A, and we define $S_{\rm I}$ to be $S \cap S_{\rm II}^{\perp}$. Obviously $S_{\rm II}$ reduces A. Hence $S_{\rm I}$ is the intersection of two invariant subspaces of A, so it is invariant under A. The irreducibility of $S_{\rm I}$ is immediate from the definition of $S_{\rm II}$, so the decomposition $S = S_{\rm II} \oplus S_{\rm I}$ has the required form. If $S = S'_{\rm II} \oplus S'_{\rm I}$ is another such decomposition, then $S'_{\rm II} \subset S_{\rm II}$. Therefore $S_{\rm II} \ominus S'_{\rm II}$ is a reducing subspace of A contained in $S'_{\rm I}$, so it must be trivial. We conclude that $S'_{\rm II} = S_{\rm II}$; in other words, the decomposition is unique.

For the case of a unitary operator we can describe the subspaces $S_{\rm II}$ and $S_{\rm I}$ more precisely. Suppose that U is a unitary operator and that S is an invariant subspace of U. The restriction of U to $S_{\rm II}$ is then unitary, so we have $S_{\rm II} = US_{\rm II} \subset US$. In fact, if n is any positive integer, then $S_{\rm II} = U^nS_{\rm II} \subset U^nS$. Hence $S_{\rm II} \subset \bigcap_{n=1}^{\infty} U^nS$. But the subspace $\bigcap_{n=1}^{\infty} U^nS$ is clearly invariant under U and U^* $(= U^{-1})$; in other words, it reduces U. Since $S_{\rm II}$ is the largest reducing subspace of U contained in S, we must have $S_{\rm II} = \bigcap_{n=1}^{\infty} U^nS$.

To analyze $S_{\rm I}$ we introduce the subspace $K = S \ominus US$, so that $S = K \oplus US$. Substituting the latter equation into itself, we find that

$$S = K \oplus U(K \oplus US) = K \oplus UK \oplus U^2S.$$

Making the substitution $S = K \oplus US$ on the right side of the last equation, we obtain $S = K \oplus UK \oplus U^2K \oplus U^3S$. This procedure can obviously be repeated; iteration yields

$$S = \left(\sum_{j=0}^{n-1} \oplus U^jK \right) \oplus U^nS, \qquad n = 1, 2, 3, \cdots. \tag{2}$$

It follows in particular that the subspaces U^mK and U^nK are orthogonal for $m \neq n$.

Because the orthogonal complement of the span of a family of subspaces is the intersection of the orthogonal complements of the subspaces in the family, we conclude from (2) that $\bigcap_{n=1}^{\infty} U^nS$ is the orthogonal complement in S of $\sum_{n=0}^{\infty} \oplus U^nK$. We have already identified the former subspace as $S_{\rm II}$, so the latter subspace must be $S_{\rm I}$.

The preceding considerations prompt a definition: We shall call a subspace L a *wandering subspace of* U if U^nL is orthogonal to L for all positive integers n. As the reader will easily verify, when this happens the subspaces U^mL and U^nL are orthogonal for all pairs of distinct integers m, n; moreover $\Sigma_{n=0}^{\infty} \oplus U^nL$ is an irreducible invariant subspace of U. The latter subspace is obviously the smallest invariant subspace of U containing L; in other words, it is the invariant subspace of U *generated* by L. We can thus restate the conclusion arrived at in the preceding paragraph as follows: *Every irreducible invariant subspace of* U *is generated by a wandering subspace of* U.

A nonzero vector is called a *wandering vector of* U if the one-dimensional subspace it spans is a wandering subspace of U. If e is a unit wandering vector of U and S_e is the invariant subspace of U generated by e, then the vectors $U^ne, n = 0, 1, 2, \cdots$, form an orthonormal basis for S_e. Corresponding to this basis there is a natural isometry of S_e onto l^2, namely, the isometry that sends the vector $\Sigma_{n=0}^{\infty} c_n U^n e$ onto the sequence $(c_0, c_1, c_2, \cdots)$. Obviously, this isometry transforms the operator $U|S_e$ into the unilateral shift operator mentioned in §1. Hence the restriction of U to the invariant subspace generated by any one-dimensional wandering subspace is unitarily equivalent to the unilateral shift. More generally, the restriction of U to an invariant subspace generated by a wandering subspace of dimension p is unitarily equivalent to the direct sum of p copies of the unilateral shift.

The above discussion, which is patterned after a paper of Halmos **[14]**, is based on extremely elementary properties of Hilbert spaces and unitary operators. Nevertheless, as we shall soon see, our conclusions have some striking analytic consequences. We consider a special class of unitary operators and we examine the meaning of our results for these operators. Let m be a finite positive Borel measure on the unit circle in the complex plane, and let U be the operator of multiplication by z on $L^2(m)$. We ask: What are the wandering vectors and wandering subspaces of U? For which measures m does U have a nonreducing invariant subspace?

Let m_0 denote Lebesgue measure on the unit circle, normalized so as to have total mass 1. The following observation is the key: *A unit vector* e *in* $L^2(m)$ *is a wandering vector of* U *if and only if* $|e|^2\, dm = dm_0$. In fact, e is a wandering vector of U if and only if

$$0 = (U^n e, e) = \int z^n |e(z)|^2\, dm(z), \qquad n = 1, 2, 3, \cdots.$$

This is equivalent to the condition

(3) $$\int z^n |e(z)|^2 \, dm(z) = \int z^n \, dm_0(z), \qquad n = 0, \pm 1, \pm 2, \cdots .$$

Since the linear span of the integral powers of z is uniformly dense in the space of continuous functions on the unit circle, condition (3) holds if and only if $|e|^2 \, dm = dm_0$.

The operator U has a nonreducing invariant subspace if and only if it has a wandering vector. From the preceding observation, therefore, we see that U *has a nonreducing invariant subspace if and only if* m_0 *is absolutely continuous with respect to* m.

By Theorem 2 (§4), a necessary and sufficient condition for U to have no nonreducing invariant subspace is that the polynomials be weak-star dense in $L^\infty(m)$. Combined with the last result, this gives the following approximation theorem: *The polynomials are weak-star dense in* $L^\infty(m)$ *if and only if* m_0 *is not absolutely continuous with respect to* m.

The wandering subspaces of U, under our present assumptions, are all of dimension one. In fact, let L be a wandering subspace of U, and let e be a unit vector in L. Let f be any vector in L which is orthogonal to e. Then $|e|^2 \, dm = dm_0$, and $|f|^2 \, dm = c^2 dm_0$, where c is the norm of f. We must show that $c = 0$, or, equivalently, that $f = 0$ almost everywhere with respect to m_0. From the definition of a wandering subspace it is immediate that $(U^n e, f) = 0$ for all integers n; in other words

$$\int z^n e(z) \overline{f(z)} \, dm(z) = 0, \qquad n = 0, \pm 1, \pm 2, \cdots .$$

Hence $e\bar{f} = 0$ almost everywhere with respect to m. Since $|e|^2 \, dm = dm_0$, the function e vanishes almost nowhere with respect to m_0, so f must vanish almost everywhere with respect to m_0, as desired.

The above results lead to a simple proof of the following theorem of F. and M. Riesz **[32]**.

THEOREM 4. *Let* ν *be a nonzero, finite, complex Borel measure on the unit circle such that*

(4) $$\int z^n \, d\nu(z) = 0, \qquad n = 1, 2, 3, \cdots .$$

Then ν *is mutually absolutely continuous with respect to* m_0.

Measures ν satisfying (4) are sometimes called *analytic measures* because their Fourier series are of analytic type, that is, they involve only nonnegative powers of z. (The quantity on the left side of (4) is the $-n$th Fourier coefficient of ν.)

To prove the Riesz theorem we form the measure $m = |\nu|$ (the total variation of ν), and, as above, we let U denote the operator on $L^2(m)$ of multiplication by z. Let S be the smallest invariant subspace of U containing the function $d\nu/dm$. Relation (4) says that the function $h(z) = \bar{z}$ is orthogonal to S. If S reduced U then $U^n h$ would be orthogonal to S for all integers n, which would mean that $\int z^n d\nu(z) = 0$ for all integers n. The latter is not the case because ν is not the zero measure, and therefore S does not reduce U. Since U has a nonreducing invariant subspace we may conclude that m_0 is absolutely continuous with respect to m. Hence m_0 is absolutely continuous with respect to ν, and one half of the theorem is proved.

To prove the other half of the theorem we note that the invariant subspace S is actually irreducible. In fact, suppose the function g in S lies in a reducing subspace of U which is contained entirely in S. Then $U^n g$ is in S for all integers n, so that

$$\begin{aligned} 0 &= (U^n g, h) \\ &= \int z^{n+1} g(z)\, dm(z), \end{aligned} \qquad n = 0, \pm 1, \pm 2, \cdots .$$

This implies that $g = 0$, which establishes the irreducibility of S.

Let S_0 be the set of functions f in $L^2(m)$ such that the measure $f dm$ is absolutely continuous with respect to m_0. Then S_0 is a reducing subspace of U and it contains all wandering vectors of U. Since S is irreducible it is generated by a wandering vector, and therefore $S \subset S_0$. In particular the function $d\nu/dm$ belongs to S_0, which means that ν is absolutely continuous with respect to m_0. The proof of the theorem is complete.

The preceding proof was discovered (but never published) by David Lowdenslager. The proof of the Riesz brothers used methods from function theory and was based on completely different ideas. Originally the Riesz theorem was motivated by a question concerning the boundary behavior of the conformal maps from the unit disk onto the interiors of rectifiable Jordan curves. The theorem implies that both the boundary function of such a map and the boundary function of its inverse are absolutely continuous with respect to arc length.

The Riesz theorem plays a crucial role in that portion of function theory in the unit disk centering around the notion of Hardy spaces **[17]**, **[21]**, **[35]** (see also §6). Attempts to extend the theory of Hardy spaces to other settings have led to generalizations of the Riesz theorem. The proof of Lowdenslager was in fact a by-product of work of this character by Helson and Lowdenslager [**18**] in which an extension of the Riesz theorem to measures on certain compact abelian groups was obtained. It has recently been recognized that the Riesz theorem is merely a special case of a very general result about measures orthogonal to function algebras [**11**, p. 44].

To conclude this section I wish to mention the connection between invariant subspaces of unitary operators and invariant subspaces of one-parameter semigroups of unitary operators. Let $\{W_t\}_{-\infty<t<\infty}$ be a strongly continuous one-parameter unitary group. By Stone's theorem, there exists a spectral measure E on the real line which provides the representation

$$W_t = \int_{-\infty}^{\infty} e^{ixt}\, dE(x), \qquad -\infty < t < \infty.$$

Let T be the (possibly unbounded) selfadjoint operator $\int_{-\infty}^{\infty} x\, dE(x)$, so that iT is the infinitesimal generator of the group $\{W_t\}$ (i.e., $W_t = e^{itT}$). Let U be the Cayley transform of T; that is, $U = (T - i)(T + i)^{-1}$. The operator U is unitary, and it does not have 1 as an eigenvalue. The following proposition reduces the study of the invariant subspaces of the semigroup $\{W_t\}_{t\geqslant 0}$ to the study of the invariant subspaces of U.

PROPOSITION 3. *A subspace is invariant under* W_t *for all* $t \geqslant 0$ *if and only if it is invariant under* U.

Let $\mathcal{A}$ be the weakly closed operator algebra generated by the semigroup $\{W_t\}_{t\geqslant 0}$, and let $\mathcal{B}$ be the weakly closed operator algebra generated by U. We shall prove the proposition by showing that $\mathcal{A} = \mathcal{B}$ (see Proposition 1, §2). The proof, while it may appear *ad hoc* to some readers, should seem natural to those who are familiar with Fourier and Laplace transforms.

Let the operator A be defined by

$$iA = \int_0^{\infty} e^{-t} W_t\, dt$$

(in other words, iA is the value at $z = 1$ of the Laplace transform of the semigroup $\{W_t\}_{t\geqslant 0}$). The right side exists as a Riemann integral in the strong operator topology. The operator iA is a strong limit of Riemann sums of the integrand, so A belongs to $\mathcal{A}$. For any vectors f and g we have

$$(Af, g) = -i\int_0^\infty e^{-t}(W_t f, g)\,dt = -i\int_0^\infty e^{-t}\left[\int_{-\infty}^\infty e^{ixt}\,d(E(x)f, g)\right]dt$$

$$= -i\int_{-\infty}^\infty \left[\int_0^\infty e^{t(ix-1)}dt\right] d(E(x)f, g)$$

$$= \int_{-\infty}^\infty (x+i)^{-1}\,d(E(x)f, g) = ((T+i)^{-1}f, g)$$

(the application of Fubini's theorem is easily justified). Hence $A = (T+i)^{-1}$. Since $U = 1 - 2i(T+i)^{-1}$, it follows that U is in A. Thus $\mathcal{B} \subset \mathsf{A}$.

To obtain the reverse inclusion consider, for a fixed $t > 0$, the functions $\varphi_r(z) = \exp\,[t(z+r)/(z-r)]$, $r \geqslant 1$, $|z| = 1$. If $r > 1$ then φ_r is the restriction to the unit circle of a function which is analytic in a neighborhood of the closed unit disk, so that φ_r can be uniformly approximated on the unit circle by polynomials. The corresponding polynomials in U will approximate $\varphi_r(U)$ in the uniform operator topology, and thus $\varphi_r(U)$ belongs to $\mathcal{B}$ for $r > 1$. As $r \longrightarrow 1$ the functions φ_r remain uniformly bounded and they converge to φ_1 at every point of the unit circle except $z = 1$. Since 1 is not an eigenvalue of U it is not a mass point for the spectral measure of U, so the dominated convergence theorem can be applied to show that $\varphi_r(U) \longrightarrow \varphi_1(U) = W_t$ in the strong operator topology. Thus W_t is in $\mathcal{B}$, and the inclusion $\mathsf{A} \subset \mathcal{B}$ follows. The proposition is proved.

6. **The bilateral shift.** In this section we let L^2 denote the space $L^2(m_0)$, where m_0, as in the last section, is normalized Lebesgue measure on the unit circle. The operator of multiplication by z on L^2 will be denoted by U_0. Of course, the results of §5 apply to U_0. We shall examine more closely what these results say about U_0 and obtain a number of consequences.

The functions $e_n(z) = z^n$, $n = 0, \pm 1, \pm 2, \cdots$, form an orthonormal basis for L^2. The operator U_0 acts by shifting this basis; that is to say, $U_0 e_n = e_{n+1}$ for all n. For this reason U_0 is called the *bilateral shift operator*. One invariant subspace of U_0 is the subspace spanned by the basis vectors e_n with $n \geqslant 0$ (in other words, the invariant subspace generated by the wandering vector e_0). We denote this subspace by H^2. The restriction operator $U_0|H^2$ is (to within unitary equivalence) the unilateral shift operator of Beurling mentioned in §1.

Suppose S is a nonreducing invariant subspace of U_0. By the results obtained in §5, there is a decomposition $S = S_{\mathrm{II}} \oplus S_{\mathrm{I}}$, where S_{II} is a reducing subspace of U_0 and S_{I} is an irreducible invariant subspace of U_0.

Moreover S_{I} is generated by a unit wandering vector w of U_0, and the vector w is uniquely determined by S_{I} to within a multiplicative constant of unit modulus. As we noted in §5, the function w satisfies $|w|^2\, dm_0 = dm_0$; in other words, $|w| = 1$ almost everywhere. From this it is clear that $S_{\mathrm{I}} = wH^2$ (the set of products wh with h in H^2). The smallest reducing subspace of U_0 containing w is obviously wL^2, in other words, L^2. This subspace must be orthogonal to S_{II}, and therefore S_{II} is trivial.

The following theorem summarizes the preceding observations. As is customary, we shall call a function in L^∞ $(= L^\infty(m_0))$ which has unit modulus almost everywhere a *unitary function.*

THEOREM 5. *Every invariant subspace of U_0 either reduces U_0 or is irreducible. The nonreducing invariant subspaces of U_0 are the subspaces $S = wH^2$ with w a unitary function in L^∞. The subspace S determines the function w to within a multiplicative constant of unit modulus.*

A unitary function which belongs to H^2 is called, following Beurling, an *inner function.* Because H^2 is an irreducible invariant subspace of U_0, the following conclusion is immediate.

COROLLARY. *The nontrivial invariant subspaces of U_0 contained in H^2 are the subspaces φH^2 with φ an inner function.*

Theorem 5 is due to Helson and Lowdenslager **[19]** and, independently, to Halmos **[14]**. I have followed the treatment of Halmos and derived the theorem from the general facts about unitary operators obtained in §5. An extremely simple direct proof of Theorem 5 due to Srinivasan (and based on the proof of Helson and Lowdenslager) can be found in [**17**, Chapter II].

The corollary is the result of Beurling **[4]** mentioned in §1. Beurling's methods were from the theory of analytic functions. The analysis given in his paper, which goes substantially beyond the above corollary, depends upon a natural representation of H^2 as a space of analytic functions. I now describe this representation.

Each function f in H^2 has an expansion $f = \sum_{n=0}^\infty c_n e_n$, where the coefficients c_n form a square summable sequence. We associate with f the analytic function in the open unit disk whose power series is $\sum_{n=0}^\infty c_n z^n$; this series has radius of convergence at least 1, because $c_n \longrightarrow 0$. We shall think of the function represented by the power series as an extension of f into the unit disk; we accordingly denote its value at z by $f(z)$. The reader will easily verify that the extended f is given in terms of the original f by Cauchy's formula:

$$f(z) = \frac{1}{2\pi i} \int_{|\zeta|=1} \frac{f(\zeta)}{\zeta - z}\, d\zeta. \tag{5}$$

The above extension procedure respects the linear structure of H^2; in other words, the extension of a linear combination of two H^2 functions is the corresponding linear combination of the separate extensions. The extension procedure thus enables us to identify H^2, as a linear space, with the space of analytic functions in the unit disk whose Taylor coefficients at the origin are square summable. The latter space is the one with which Beurling worked. The unilateral shift corresponds to the operator on it of multiplication by z.

A function f in H^2 is the boundary function, in a L^2-sense, of its analytic extension. To be precise, let the functions f_r $(0 < r < 1)$ on the unit circle be defined by $f_r(z) = f(rz)$. Then the functions f_r belong to H^2, and a simple estimate shows that they converge to f in L^2-norm as r tends to 1. There is another sense in which f is the boundary function of its analytic extension. Namely, by a famous theorem of Fatou [**21**, p. 34], the analytic extension of f converges to f along almost every radius; that is to say $f(z) = \lim_{r \to 1-} f(rz)$ for almost every point z on the unit circle. The theorem of Fatou is difficult to prove. The following weaker result, which is a consequence of the L^2-convergence of the functions f_r to f, will suffice for our immediate purposes: There exists a sequence $\{r_n\}_{n=1}^{\infty}$ $(0 < r_n < 1)$ tending to 1 such that $f(z) = \lim_{n \to \infty} f(r_n z)$ for almost every z on the unit circle.

By the preceding result, if the analytic extension of an H^2 function is bounded in the unit disk, then the function itself is essentially bounded. We let H^∞ denote the space of essentially bounded functions in H^2. The space H^2 is invariant under multiplication by the functions in H^∞; in other words, if f is in H^2 and h is in H^∞, then hf is in H^2. To prove this, choose a sequence $\{p_n\}_{n=1}^{\infty}$ of polynomials converging to f in L^2-norm; for example, if $f = \sum_{n=0}^{\infty} c_n e_n$, we can take $p_n = \sum_{k=0}^{n} c_k e_k$. Because H^2 is invariant under U_0 it contains each product $hp_n = p_n(U_0)h$. Because h is bounded the sequence $\{hp_n\}$ converges in L^2-norm to hf, so hf belongs to H^2, as asserted. Moreover, if $|z| < 1$, the equality $(hp_n)(z) = h(z)p_n(z)$ is easily verified. The Cauchy formula (5) implies that evaluation at z is a bounded linear functional on H^2. We can thus pass to the limit and obtain $(hf)(z) = h(z)f(z)$. Our process of analytic extension, therefore, preserves multiplication, at least when one of the factors is bounded.

We have observed that H^∞ contains every H^2 function whose analytic extension is bounded. By using a trick of Landau we can show that, conversely, the analytic extension of any H^∞ function is bounded. In fact, let h be in H^∞, and fix a point z in the unit disk. If n is any positive integer then, by what we proved above, h^n is in H^∞ and $(h^n)(z) = h(z)^n$. Hence

$$h(z)^n = \frac{1}{2\pi i}\int_{|\zeta|=1}\frac{h(\zeta)^n}{\zeta - z}\,d\zeta.$$

This implies that $|h(z)|^n \leqslant (1-|z|)^{-1}\|h\|_\infty^n$; in other words, $|h(z)| \leqslant (1-|z|)^{-1/n}\|h\|_\infty$. Letting n approach ∞, we obtain the inequality $|h(z)| \leqslant \|h\|_\infty$.

It remains to add that every bounded analytic function in the unit disk has square summable Taylor coefficients and so is the analytic extension of an H^∞ function. This follows from the equality

$$\sum_{n=0}^{\infty}|c_n|^2 r^{2n} = \frac{1}{2\pi}\int_0^{2\pi}\left|\sum_{n=0}^{\infty}c_n r^n e^{in\theta}\right|^2 d\theta, \qquad 0<r<1, \tag{6}$$

which holds for any function $\Sigma_0^\infty c_n z^n$ analytic in the unit disk. We therefore see that H^∞ can be identified with the space of bounded analytic functions in the unit disk. This identification is an algebraic isomorphism; that is, it is one-to-one and it preserves addition and multiplication.

The inner functions are the functions in H^∞ having unit modulus almost everywhere. They form a semigroup under multiplication, and divisibility in this semigroup corresponds to an inclusion relation between invariant subspaces. More precisely, if φ and ψ are inner functions, one says that φ divides ψ provided there is a factorization $\psi = \varphi\psi_1$ with ψ_1 an inner function. This happens, obviously, if and only if the invariant subspace φH^2 contains the invariant subspace ψH^2. By the uniqueness statement in Theorem 5, two inner functions divide each other if and only if they are constant multiples of each other.

We now use the above analytic representation to gain further insight into the invariant subspaces of H^2 and into the structure of inner functions. To begin we consider a very simple kind of invariant subspace. Fix a point a in the unit disk, and let $S(a)$ denote the set of functions f in H^2 such that $f(a) = 0$. We noted above that evaluation at a is a bounded linear functional on H^2. Thus $S(a)$ is a subspace of H^2 of codimension one. It is obviously invariant under U_0, so it is determined by an inner function. What is this inner function?

The answer is obvious in case $a = 0$; the desired inner function is then the function z (or λz with λ any constant of unit modulus). We shall thus suppose that $a \neq 0$. For the sake of definiteness we seek the inner function b_a

such that $S(a) = b_a H^2$ and such that $b_a(0)$ is real and positive. There is precisely one such function, because the equality $S(a) = b_a H^2$ determines b_a to within a multiplicative constant of unit modulus. (For completeness, we let b_0 be the function z.)

To find b_a we note that the orthogonal complement of $S(a)$ in H^2 is spanned by the function in H^2 that induces the functional of evaluation at a. We denote this function by K_a; it is usually called the *kernel function* in H^2 for the point a. It is easy to find K_a explicitly; we have, for $n \geqslant 0$,

$$(K_a, e_n) = \overline{e_n(a)} = \bar{a}^n,$$

and thus $K_a = \Sigma_{n=0}^{\infty} \bar{a}^n e_n$. The series $\Sigma_0^{\infty} \bar{a}^n e_n(z) = \Sigma_0^{\infty} \bar{a}^n z^n$ is a geometric series with sum $1/(1 - \bar{a}z)$. The convergence is uniform in the closed unit disk, and therefore $K_a(z) = 1/(1 - \bar{a}z)$ $(|z| \leqslant 1)$. (The reader will easily convince himself that the expression $f(a) = (f, K_a)$ $(f \in H^2)$ is just another form of the Cauchy formula (5).)

Now let f be any function in $S(a)$, say $f = b_a g$ with g in H^2. We observe that

$$\begin{aligned}(f, b_a) = (b_a g, b_a) &= \int |b_a|^2 g \, dm_0 \\ &= \int g \, dm_0 = g(0) = f(0)/b_a(0).\end{aligned}$$

Thus $(f, b_a(0) b_a) = (f, K_0)$ for all f in $S(a)$. It follows that $b_a(0) b_a$ is the projection onto $S(a)$ of K_0. This projection is equal to

$$K_0 - \frac{(K_0, K_a)}{\|K_a\|^2} K_a,$$

in other words (since K_0 is the constant function 1), to the function

$$1 - \frac{1 - |a|^2}{1 - \bar{a}z} = \frac{\bar{a}(a - z)}{1 - \bar{a}z}.$$

The value of the latter function at $z = 0$ must be $b_a(0)^2$. Hence $b_a(0) = |a|$, and we obtain

$$b_a(z) = \frac{\bar{a}}{|a|} \cdot \frac{a - z}{1 - \bar{a}z}. \tag{7}$$

The function b_a is called the *Blaschke factor* for the point a. It is the unique conformal map of the unit disk onto itself that takes a to 0 and 0 to the positive real axis. Blaschke factors turn up in many different problems in function theory, and the reader who has encountered them before possibly arrived at formula (7) without having to go through the above reasoning.

To pursue these ideas further, let a_1 and a_2 be two points in the unit disk, and let $S(a_1, a_2)$ be the set of functions f in H^2 such that $f(a_1) = f(a_2) = 0$. Then $S(a_1, a_2)$ is an invariant subspace of U_0, so it is equal to BH^2 for some inner function B. To find B we note that B divides the inner function $b_{a_1}b_{a_2}$, because the latter function belongs to $S(a_1, a_2)$. On the other hand, since B belongs to $S(a_1)$ it is divisible by b_{a_1}, in other words, the function B/b_{a_1} is an inner function. Moreover B/b_{a_1} belongs to $S(a_2)$, so it is divisible by b_{a_2}. Hence B is divisible by $b_{a_1}b_{a_2}$. Therefore B is a constant multiple of $b_{a_1}b_{a_2}$, in other words, we may take $B = b_{a_1}b_{a_2}$.

The above reasoning can obviously be extended from two points to finitely many points. Let $\alpha = (a_1, a_2, \cdots, a_p)$ be a finite sequence of points in the unit disk, and let $S(\alpha)$ be the set of functions f in H^2 such that $f(a_n) = 0$, $n = 1, \cdots, p$. We allow repetitions in the sequence α with the understanding that at a repeated point, the functions in $S(\alpha)$ are required to vanish with multiplicities at least as great as the number of repetitions. It is easily seen that $S(\alpha)$ is a subspace of H^2. Using the above reasoning, one can show that it is the invariant subspace corresponding to the inner function $\Pi_{n=1}^{p} b_{a_n}$. A function of the latter form is called a *finite Blaschke product.*

With a little more work one can obtain an analogous result for infinite zero sets. Let $\alpha = \{a_n\}_{n=1}^{\infty}$ be an infinite sequence of points in the unit disk, and let $S(\alpha)$ be the subspace of functions in H^2 that vanish at each point of α (with the appropriate multiplicities at repeated points). The following proposition describes the inner function corresponding to $S(\alpha)$.

PROPOSITION 4. *The subspace* $S(\alpha)$ *is nontrivial if and only if*

$$\sum_{n=1}^{\infty} (1 - |a_n|) < \infty. \tag{8}$$

When (8) *holds, the infinite product*

$$\prod_{n=1}^{\infty} b_{a_n} \tag{9}$$

converges in L^2*-norm to an inner function* B, *and* $S(\alpha) = BH^2$.

An inner function which can be represented by a product of the form (9) is called an *infinite Blaschke product.*

I shall only sketch the proof of Proposition 4. The omitted details are all very elementary and the interested reader should have no trouble supplying them.

We restrict our attention to the case where 0 does not appear in the sequence α; the general case is easily reduced to this one. For $p = 1, 2, \cdots$ let B_p be the finite Blaschke product $\Pi_{n=1}^{p} b_{a_n}$. If p and k are any positive integers then

$$\begin{aligned}
\|B_{p+k} - B_p\|^2 &= \int |B_{p+k} - B_p|^2\, dm_0 \\
&= \int |B_{p+k}|^2\, dm_0 + \int |B_p|^2\, dm_0 - 2\,\mathrm{Re} \int B_{p+k}\bar{B}_p\, dm_0 \\
&= 2 - 2\,\mathrm{Re} \int \left[\prod_{n=p+1}^{p+k} b_{a_n} \right] dm_0 \\
&= 2\left[1 - \prod_{n=p+1}^{p+k} b_{a_n}(0) \right] = 2\left[1 - \prod_{n=p+1}^{p+k} |a_n| \right].
\end{aligned}$$

Thus the L^2-convergence of the product (9) is equivalent to the condition $\lim_{p,k\to\infty} \Pi_{n=p+1}^{p+k} |a_n| = 1$, which, in turn, is equivalent to the condition

$$\prod_{n=1}^{\infty} |a_n| > 0. \tag{10}$$

The latter condition is of course equivalent to (8).

If the product (9) converges in L^2-norm then its limit is clearly an inner function in $S(\alpha)$, so in particular $S(\alpha)$ is nontrivial. Suppose, conversely, that $S(\alpha)$ is nontrivial. Then it has the form $S(\alpha) = \varphi H^2$ for some inner function φ. Since 0 does not appear in the sequence α we have $\varphi(0) \neq 0$. The function φ belongs to $S(a_1, a_2, \cdots, a_p)$ for each p, and therefore there are factorizations $\varphi = B_p\varphi_p$ with φ_p inner. Since $|\varphi_p(0)| \leqslant 1$ we have

$$|\varphi(0)| \leqslant |B_p(0)| = \prod_{n=1}^{p} |a_n|,$$

and therefore (10) holds. Thus the product (9) converges in L^2; let B be its limit. As noted above, B belongs to $S(\alpha)$, so φ divides B. On the other hand, the sequence $\{\varphi\bar{B}_p\}$ converges in L^2 to $\varphi\bar{B}$. Since $\varphi\bar{B}_p = \varphi_p$ is in H^2 for each p, the function $\varphi\bar{B} = \varphi/B$ is also in H^2, so that B divides φ. Hence B is a constant multiple of φ, and we have $S(\alpha) = BH^2$, as desired. With this, Proposition 4 is established.

In particular, Proposition 4 identifies the zero sets of the analytic functions in the unit disk with square summable Taylor coefficients; it says that these zero sets are precisely the sequences $\{a_n\}$ satisfying (8). For a general analytic function in the unit disk, of course, all one can say concerning its zeros is that they cluster only on the unit circle (unless the function is identically zero).

The question now arises whether there are inner functions other than the Blaschke products. In order to answer this question, we consider an arbitrary inner function φ. If B is the Blaschke product corresponding to the zeros of φ, then φ lies in BH^2 by Proposition 4, so there is a factorization $\varphi = B\psi$ where ψ is an inner function without zeros. (If φ has no zeros we set $B \equiv 1$.) The function ψ has an analytic logarithm in the unit disk, which we denote by $\log \psi$ (it is unique to within addition of an integral multiple of $2\pi i$). Since $|\psi(z)| \leqslant 1$ in the unit disk, the function $-\log \psi$ has a nonnegative real part. There is a classical theorem of Herglotz **[21**, p. 34] concerning representation of analytic functions in the unit disk with nonnegative real parts; it tells us that there is a finite nonnegative Borel measure μ on the unit circle, and a real constant c, such that

$$(11) \qquad \log \psi(z) = ic - \int \frac{\zeta + z}{\zeta - z}\, d\mu(\zeta), \qquad |z| < 1.$$

Moreover, by the theorem of Fatou mentioned earlier, the radial limits of the real part of $-\log \psi$ exist almost everywhere (modulo m_0) and equal the Radon-Nikodym derivative of μ with respect to m_0. Now because ψ is an inner function we have $\lim_{r \to 1} |\psi(rz)| = |\psi(z)| = 1$ for almost every point z on the unit circle; in other words, the radial limits of the real part of $\log \psi$ vanish almost everywhere. Hence $d\mu/dm_0 = 0$ almost everywhere; that is to say, μ must be singular with respect to m_0.

Exponentiating (11), we obtain the expression

$$(12) \qquad \psi(z) = \lambda \exp \left[-\int \frac{\zeta + z}{\zeta - z}\, d\mu(\zeta) \right], \qquad |z| < 1,$$

where $\lambda = \exp(ic)$. As the above remarks indicate, any function ψ of the form (12), where λ is a constant of unit modulus and μ is a nonnegative singular measure on the unit circle, is an inner function. Inner functions of this form are called *singular functions*. The reasoning above shows that every inner function can be factored into the product of a singular inner function and a Blaschke product (where the latter factor may reduce to the constant 1).

An inner function, therefore, is determined to within a multiplicative constant by a zero set in the unit disk (a finite sequence or an infinite sequence satisfying (8)) and a finite nonnegative singular Borel measure on the unit circle. In terms of these parameters, divisibility has the following interpretation: If φ_1 and φ_2 are inner functions, then φ_1 divides φ_2 if and only if the zero set of φ_2 contains the zero set of φ_1 and the singular measure of φ_2 dominates the singular measure of φ_1. With this information on the structure of inner functions, we obtain a very clear picture of the structure of the unilateral shift's invariant subspace lattice. (We shall not attempt to prove here the above divisibility criterion. The part concerning zero sets is obvious, but the part concerning singular measures lies deeper.)

The spaces H^2 and H^∞ with which we have been working are two of a family of spaces, called Hardy spaces, around which an extensive theory has developed. For $1 \leqslant p \leqslant \infty$, the Hardy space H^p is defined to be the subspace of L^p consisting of those functions f in L^p satisfying

$$\int z^n f(z)\, dm_0(z) = 0, \qquad n = 1, 2, \cdots .$$

Thus, each function f in H^p has a Fourier series of analytic type (that is, involving only nonnegative powers of z), and, just as in the case $p = 2$, we can extend f analytically into the unit disk by reinterpreting the Fourier series of f as a power series. For finite p the analytic extension of f satisfies the growth condition

$$\sup_{r<1} \int_0^{2\pi} |f(re^{i\theta})|^p \, d\theta < \infty,$$

and this condition characterizes the analytic extensions of H^p functions. (For $p = 2$ the last assertion follows from (6).) The original function f is the boundary function of its analytic extension in an L^p-sense as well as in the sense of radial limits. The factorization of inner functions described above is a special case of a theorem on factorization of general H^p functions.

Hardy spaces were studied originally by Hardy, Littlewood, F. Riesz, M. Riesz, Szegö, Smirnov, and many other analysts. Interest in the subject mushroomed after Beurling's signal paper on the unilateral shift, and the last twenty years have seen an enormous volume of work, much of it motivated by functional analytic considerations. The book of Hoffman **[21]** has a nice account of the theory along with many references to the literature.

The above described results on the bilateral and unilateral shifts have been extended in several directions. One extension, due to Lax [**24**] and Halmos [**14**], concerns shifts of higher multiplicities. If ν is a cardinal number (which, to avoid complications, we assume is $\leqslant \aleph_0$), then the *bilateral shift of multiplicity* ν is, by definition, the direct sum of ν copies of U_0. This operator is conveniently represented as multiplication by z on $L^2(m_0; \mathcal{V})$, where $\mathcal{V}$ is a Hilbert space of dimension ν (see §3). Lax and Halmos show that the irreducible invariant subspaces of a multiple shift have a description analogous to the one given in Theorem 5 for the shift of unit multiplicity. In their description, the unitary function that appears in Theorem 5 is replaced by a certain kind of operator-valued function. The reader will find further information on this subject in the book of Helson [**17**].

Other extensions of Theorem 5 occur in connection with extensions of the theory of Hardy spaces to more general settings; see for example [**11**, p. 132] and [**38**].

With an eye to an application in §8, I shall end this section by applying the results at the end of the last section to the one-parameter unitary semigroup associated with U_0. Let T_0 be the Cayley transform of U_0 and let $\{W_t\}_{-\infty<t<\infty}$ be the one-parameter unitary group whose infinitesimal generator is iT_0. Clearly, T_0 is the transformation on L^2 of multiplication by $i(1+z)/(1-z)$ (on the domain of all f in L^2 such that $(1+z)f/(1-z)$ is in L^2). Thus, W_t is the operator on L^2 of multiplication by $\exp[t(1+z)/(1-z)]$. By Proposition 3, the semigroup $\{W_t\}_{t\geqslant 0}$ has the same invariant subspaces as U_0.

The form of T_0 suggests that we apply a change of variables. Consider the map of the unit circle onto the real line that sends z onto $i(1+z)/(1-z)$. This map transforms normalized Lebesgue measure on the circle into the measure $\pi^{-1}(1+x^2)^{-1}\,dx$. Its inverse sends x onto $(x-i)/(x+i)$. Hence we can define an isometry J_0 of L^2 onto $L^2(-\infty,\infty)$ by

$$(J_0 f)(x) = \pi^{-\frac{1}{2}}(x+i)^{-1} f\left(\frac{x-i}{x+i}\right).$$

The isometry J_0 transforms T_0 into the operator of multiplication by x and W_t into the operator of multiplication by $\exp(itx)$.

We now follow J_0 by the Fourier transformation F. Recall that F is a unitary operator of $L^2(-\infty,\infty)$ onto itself. On a function g which belongs to both $L^2(-\infty,\infty)$ and $L^1(-\infty,\infty)$, the transformation F is defined by

$$(Fg)(x) = (2\pi)^{-\frac{1}{2}}\int_{-\infty}^{\infty} e^{-ix\xi} g(\xi)\,d\xi. \tag{13}$$

The action of F on a function not in $L^1(-\infty, \infty)$ can often be determined by expressing the function as the L^2-limit of a sequence in $L^1(-\infty, \infty)$ and applying (13) to the functions in the sequence. The action of F^{-1} on a function g belonging to both $L^2(-\infty, \infty)$ and $L^1(-\infty, \infty)$ is given by

$$(F^{-1}g)(x) = (2\pi)^{-1/2}\int_{-\infty}^{\infty} e^{ix\xi}g(\xi)\, d\xi. \tag{14}$$

A thorough discussion of the Fourier transformation can be found in [**35**, Chapter 9].

It is clear from (13) that F transforms the operator of multiplication by $\exp(itx)$ into the operator of translation by t (that is, the operator that sends the function $h(x)$ onto the function $h(x-t)$). Hence, the composite map $J = FJ_0$ transforms the group $\{W_t\}$ into the group of translations on $L^2(-\infty, \infty)$. A subspace of $L^2(-\infty, \infty)$ is invariant under all right translations if and only if it is the image under J of an invariant subspace of U_0. This in conjunction with Theorem 5 describes the right-translation invariant subspaces of $L^2(-\infty, \infty)$.

One right-translation invariant subspace of $L^2(-\infty, \infty)$ is the subspace of functions that vanish almost everywhere on $(-\infty, 0)$; we may identify this subspace in the obvious way with $L^2(0, \infty)$. It turns out that $L^2(0, \infty)$ is the image under J of H^2. To prove this we first determine Je_0. Since $(J_0e_0)(x) = \pi^{-1/2}(x+i)^{-1}$, the function Je_0 is the L^2-limit as $r \to \infty$ of the functions h_r defined by

$$h_r(x) = (2\pi)^{-1/2}\int_{-r}^{r} \pi^{-1/2}(\xi + i)^{-1}e^{ix\xi}\, d\xi.$$

A standard application of the residue theorem, which I leave to the reader, shows that

$$\begin{aligned}\lim_{r\to\infty} h_r(x) &= i2^{1/2}e^{-x} && \text{for } x > 0,\\ &= 0 && \text{for } x < 0.\end{aligned}$$

Hence Je_0 is in $L^2(0, \infty)$.

Since $L^2(0, \infty)$ is right-translation invariant, it is invariant under the operator JU_0J^{-1}. Therefore it contains $(JU_0J^{-1})^nJe_0 = Je_n$ for all positive integers n. It follows that $JH^2 \subset L^2(0, \infty)$.

To prove that the last inclusion is an equality, it will be enough to show that J sends $H^{2\perp}$ into $L^2(-\infty, 0)$. (This follows because $L^2(-\infty, \infty)$ must be the direct sum of JH^2 and $JH^{2\perp}$.) Now J_0e_{-1} is the complex conjugate of $J_0 e_0$. By (13), the Fourier transform of the complex conjugate of a function g is the reflection about the origin of the complex conjugate of the Fourier transform of g. Since Je_0 is in $L^2(0, \infty)$ it follows that Je_{-1} is in $L^2(-\infty, 0)$. Because $L^2(-\infty, 0)$ is left-translation invariant it is invariant under $JU_0^{-1}J^{-1}$. Thus, by a repetition of the above reasoning, $L^2(-\infty, 0)$ contains Je_n for all negative integers n, and this implies the desired inclusion $JH^{2\perp} \subset L^2(-\infty, 0)$.

As mentioned above, the preceding observations will be needed later, in §8.

7. **Maximal subalgebras.** When Theorem 5, which describes the invariant subspaces of the bilateral shift, is combined with Theorem 2, which identifies certain reflexive operator algebras, an interesting property of H^∞ emerges. In the last section (the notations of which we retain here) we showed that H^∞ is a subalgebra of L^∞. In addition, H^∞ is closed with respect to the weak-star topology of L^∞ because it is the intersection of the kernels of the linear functionals on L^∞ induced by the functions $z^n, n = 1, 2, \cdots$.

PROPOSITION 5. *H^∞ is a maximal weak-star closed subalgebra of L^∞; that is to say, if B is a weak-star closed subalgebra of L^∞ which contains H^∞, then either $B = L^\infty$ or $B = H^\infty$.*

To prove this, we assume $B \neq L^\infty$. By Theorem 2, the algebra of multiplication operators on L^2 induced by the functions in B is reflexive. Therefore, there is some subspace S of L^2 which is invariant under multiplication by all the functions in B but not under multiplication by all the functions in L^∞. Since $B \supset H^\infty$, the subspace S is invariant under U_0. It is therefore a nonreducing invariant subspace of U_0; so, by Theorem 5, it equals wH^2 for some unitary function w. If now h is a function in B, then $hwH^2 \subset wH^2$. This implies that $hH^2 \subset H^2$, and hence that h is in H^2. Thus h is in H^∞, and we may conclude that $B = H^\infty$, as desired.

Proposition 5 can be reformulated as a statement about weak-star approximation. First we remark that the polynomials are weak-star dense in H^∞ (in fact, H^∞ is sometimes defined as the weak-star closure of the polynomials in L^∞, rather than as $H^2 \cap L^\infty$). This follows, for example, from the following result in Fourier analysis [**21**, p. 19]: If f is any function in L^∞, then the Cesàro means of the Fourier series of f (which are polynomials if f is in H^∞) converge to f in the weak-star topology of L^∞. It also follows (in less elementary fashion) from the observation that the above proof of Proposition 5, when slightly modified, shows that the weak-star closure of the polynomials in L^∞ is itself a maximal weak-star closed subalgebra of L^∞; this algebra, since it is contained in H^∞, must therefore equal H^∞.

This noted, we can restate Proposition 5 as follows: *If* f *is any function in* L^∞ *which is not in* H^∞, *then the polynomials in* f *and* z *are weak-star dense in* L^∞. In fact, the weak-star closure of the polynomials in f and z is a subalgebra of L^∞. It contains H^∞ since it contains the ordinary polynomials, and it does not equal H^∞ since it contains f. Therefore, by Proposition 5, it equals L^∞.

From Proposition 5 we can obtain another result on maximal subalgebras. Let C denote the space of continuous complex-valued functions on the unit circle, and let $A = H^\infty \cap C$. Obviously A is a uniformly closed subalgebra of C.

PROPOSITION 6. *A is a maximal uniformly closed subalgebra of C; that is to say, if B is a uniformly closed subalgebra of C which contains A, then either $B = C$ or $B = A$.*

To prove this we assume $B \neq A$. By the Hahn-Banach and Riesz representation theorems, in order to show that $B = C$ it will be enough to show that the only Borel measure on the unit circle that annihilates B is the zero measure.

Let ν be a Borel measure on the unit circle which annihilates B. Then ν annihilates A, so, in particular, it annihilates all positive powers of z. Therefore, by the F. and M. Riesz theorem, ν is absolutely continuous with respect to m_0; let $g = d\nu/dm_0$. By Proposition 5 (and the remark following its proof), B is weak-star dense in L^∞. Since $\int fg\, dm_0 = \int f\, d\nu = 0$ for all f in B, it follows that $g = 0$, and hence that $\nu = 0$, as desired.

We could have defined the algebra A as the uniform closure in C of the polynomials. The uniform density of the polynomials in A follows, for example, from a classical theorem of Fejér [**21**, p. 17] which states that the Cesàro means of the Fourier series of a function in C converge uniformly to the function; if the function is in A, these Cesàro means are polynomials. In view of this one can, as with Proposition 5, reformulate Proposition 6 as a result about approximation: *If* f *is any function in* C *which is not in* A, *then the polynomials in* f *and* z *are uniformly dense in* C. If one chooses for f the function $\bar{z}$, this reduces to the classical theorem of Weierstrass which asserts the uniform density of the trigonometric polynomials in C.

Proposition 6, which was the earliest result on maximal subalgebras, is due to Wermer [**42**]. Proposition 5, to the best of my knowledge, was discovered by Hoffman and Singer and first published in Hoffman's book [**21**, p. 194]. A systematic study of maximal subalgebras can be found in the paper [**22**] of Hoffman and Singer.

8. The Volterra operator. In this section we determine the invariant subspaces of the operator V on $L^2(0, 1)$ defined by

$$(Vf)(x) = \int_0^x f(t)\, dt.$$

This is the simplest example of a Volterra integral operator, that is, an integral operator on $L^2(0, 1)$ whose kernel belongs to L^2 of the unit square and vanishes above the main diagonal. Usually V is referred to simply as *the Volterra operator.*

The spectrum of V consists of the origin alone. There are two ways of proving this, and the reader should have no difficulty supplying the details of either proof. One way is to show directly that the spectral radius of V is zero, that is, to estimate $\|V^n\|$ $(n = 1, 2, 3, \cdots)$ and show that $\lim_{n\to\infty} \|V^n\|^{1/n} = 0$. Alternatively, one can construct explicitly the inverse of $V - \lambda$ for any nonzero scalar λ; this amounts to solving a simple first order differential equation with appropriate boundary conditions. (It is in fact the case that the spectrum of any Volterra integral operator consists of 0 alone; see [**16**, Problems 146 and 147].)

Certain invariant subspaces of V are immediately visible: If $0 \leqslant a \leqslant 1$, then the subspace of $L^2(0, 1)$ consisting of those functions that vanish almost everywhere on $(0, a)$ is obviously invariant under V. It turns out that these are the only invariant subspaces of V. The result is nontrivial; as we shall see, it is intimately related to a theorem of Titchmarsh on convolution. We prove the result here by exploiting a connection between V and the shift operator. This connection is implemented by the transformation J introduced at the end of §6.

We retain the notations of §6. The transformation J, we recall, is an isometry of L^2 onto $L^2(-\infty, \infty)$, and it sends H^2 onto $L^2(0, \infty)$. (As before, we identify $L^2(0, \infty)$ in the obvious way with a subspace of $L^2(-\infty, \infty)$. We adopt the same convention with regard to all the spaces $L^2(a, b)$, $-\infty \leqslant a \leqslant b \leqslant \infty$.)

For a any real number, let ψ_a be the unitary function on the unit circle defined by $\psi_a(z) = \exp[a(z + 1)/(z - 1)]$. We note for a positive ψ_a is a singular inner function whose corresponding singular measure is a point mass of magnitude a concentrated at $z = 1$. As we saw in §6, the isometry J transforms the operator on L^2 of multiplication by ψ_a into the operator on $L^2(-\infty, \infty)$ of translation by a. Hence J sends $\psi_a H^2$ onto $L^2(a, \infty)$, and, in particular, it sends $\psi_1 H^2$ onto $L^2(1, \infty)$. Consequently, J sends the subspace $K = H^2 \ominus \psi_1 H^2$ onto $L^2(0, 1)$. This noted, it is natural to try to determine the operator on K that is transformed into V by J.

We recall first a few facts about convolution. (Proofs can be found in [**35**, pp. 146–147 and 180–181].) If f and g are functions in $L^1(-\infty, \infty)$, then the convolution of f and g is the function $f * g$ on $(-\infty, \infty)$ defined by

$$(f * g)(x) = \int_{-\infty}^{\infty} f(t)g(x - t)\, dt. \tag{15}$$

The integral on the right exists for almost every x, and the function $f * g$ belongs to $L^1(-\infty, \infty)$. Convolution is commutative and associative; that is, $f * g = g * f$, and if h is another function in $L^1(-\infty, \infty)$ then $(f * g) * h = f * (g * h)$. The Fourier transform of $f * g$ is the product of the Fourier transforms of f and g multiplied by the constant $(2\pi)^{1/2}$. The inverse Fourier transform of $f * g$ has a similar description.

We now transform V by J^{-1}, applying successively the maps F^{-1} and J_0^{-1}. Let χ denote the characteristic function of $(0, 1)$. Then, for any f in $L^2(0, 1)$, one can think of Vf as being the projection onto $L^2(0, 1)$ of the convolution $\chi * f$. Hence F^{-1} sends V into the operator on $F^{-1}L^2(0, 1)$ of multiplication by the function $\omega = (2\pi)^{1/2}F^{-1}\chi$ followed by projection onto $F^{-1}L^2(0, 1)$. That operator, in turn, is transformed by J_0^{-1} into the operator on K of multiplication by the function $\varphi(z) = \omega(i(1 + z)/(1 - z))$ followed by projection onto K. The latter operator, which we denote by Y, is the transform of V under J^{-1}.

Using (14), we find that $\omega(x) = (e^{ix} - 1)/ix$. Hence

$$\begin{aligned} \varphi(z) &= (z - 1)(z + 1)^{-1}[\psi_1(z) - 1] \\ &= [1 - 2(z + 1)^{-1}][\psi_1(z) - 1]. \end{aligned}$$

We can accordingly write Y as the sum of two operators: the operator of multiplication by $\psi_1(z) - 1$ followed by projection onto K, and the operator of multiplication by $2(z + 1)^{-1}[1 - \psi_1(z)]$ followed by projection onto K. As the reader will easily verify, the former operator is the negative of the identity, and the latter operator is the inverse of the operator of multiplication by $(z + 1)/2$ followed by projection onto K. Hence, if we let Z denote the operator on K of multiplication by z followed by projection onto K, we have $(Y + 1)^{-1} = (Z + 1)/2$; in other words, J transforms $(Z + 1)/2$ into $(V + 1)^{-1}$ (and transforms $(1 - Z)(1 + Z)^{-1}$ into V).

Now the operators V and $(V + 1)^{-1}$ have the same invariant subspaces, because each lies in the uniformly closed operator algebra generated by the other. In fact, since the spectrum of V consists of the origin alone, the operator

$(V + 1)^{-1}$ is given by a norm convergent Neumann series in V:

$$(V + 1)^{-1} = \sum_{n=0}^{\infty} (-1)^n V^n.$$

Similarly, V is given in terms of $(V + 1)^{-1}$ by a norm convergent series:

$$V = \sum_{n=1}^{\infty} (-1)^n [(V + 1)^{-1} - 1]^n.$$

The invariant subspaces of V are therefore the images under J of the invariant subspaces of $(Z + 1)/2$, that is, of the invariant subspaces of Z.

The subspace K is the orthogonal complement in H^2 of the subspace $\psi_1 H^2$, which is invariant under U_0. Therefore K is invariant under $(U_0|H^2)^*$. The restriction to K of $(U_0|H^2)^*$ is Z^*. Hence the invariant subspaces of Z^* are precisely those invariant subspaces of $(U_0|H^2)^*$ that are contained in K, in other words, the orthogonal complements in H^2 of those invariant subspaces of U_0 lying between H^2 and $\psi_1 H^2$. By Beurling's theorem, the latter subspaces are the subspaces ψH^2 with ψ an inner function dividing ψ_1. The invariant subspaces of Z are the orthogonal complements in K of the invariant subspaces of Z^*; they are therefore the subspaces $\psi H^2 \ominus \psi_1 H^2$ with ψ an inner function dividing ψ_1. The problem of finding the invariant subspaces of V is thus reduced to the problem of finding the inner divisors of ψ_1.

If ψ is an inner function dividing ψ_1 then, because ψ_1 has no zeros in the unit disk, ψ also has no zeros in the unit disk. Hence ψ must be a singular inner function. Moreover, by the factorization criterion mentioned (but not proved) in §6, the singular measure associated with ψ must be dominated by the singular measure associated with ψ_1, that is, by the unit point measure at $z = 1$. The measure associated with ψ is therefore a point measure at $z = 1$ of mass a, where $0 \leqslant a \leqslant 1$, and ψ accordingly is a constant multiple of ψ_a. (A more elementary method of identifying the inner divisors of ψ_1 is given by Lax in [**24**, p. 173].)

The invariant subspaces of Z are therefore the subspaces $\psi_a H^2 \ominus \psi_1 H^2$, $0 \leqslant a \leqslant 1$. The image of $\psi_a H^2 \ominus \psi_1 H^2$ under J is $L^2(a, 1)$. Hence the invariant subspaces of V are the subspaces $L^2(a, 1)$, $0 \leqslant a \leqslant 1$.

The above method of finding the invariant subspaces of V is due to the author [**36**]. Brodskiĭ [**6**] and Donoghue [**9**] were the first to determine the invariant subspaces of V, the former author using the spectral theory for nonselfadjoint operators developed by Livšic, himself, and other Russian mathematicians, the latter author using the convolution theorem of Titchmarsh that I mentioned above and now state.

For f a measurable function on the line, let $\sigma(f)$ denote the largest number a such that f vanishes almost everywhere on $(-\infty, a)$ $(\sigma(f) = -\infty$ if there is no such a and $\sigma(f) = +\infty$ if $f = 0$ almost everywhere). If f and g are locally integrable functions on the line (that is, integrable on each finite interval), and if $\sigma(f) > -\infty$ and $\sigma(g) > -\infty$, then the convolution $f * g$ is well defined by (15) and is a locally integrable function. Titchmarsh's theorem reads as follows.

THEOREM 6. *Let f and g be locally integrable functions on the line such that $\sigma(f) > -\infty$ and $\sigma(g) > -\infty$. Then $\sigma(f * g) = \sigma(f) + \sigma(g)$.*

Kalisch [23] has pointed out that the reasoning of Donoghue can be reversed; that is to say, one can base a proof of the Titchmarsh theorem on the fact that the only invariant subspaces of V are the subspaces $L^2(a, 1)$. The argument is as follows.

As the inequality $\sigma(f * g) \geqslant \sigma(f) + \sigma(g)$ is trivial, we need only establish the reverse inequality. The following lemma does this for a special case to which the general case is easily reduced.

LEMMA. *Let f and g be functions in $L^2(0, b)$, where $0 < b < \infty$. Assume that $\sigma(f) = 0$ and $\sigma(f * g) \geqslant b$. Then g is the zero function.*

Applying the change of variables $x \longrightarrow x/b$, we may assume without loss of generality that $b = 1$. Since f and g are square integrable, the function $f * g$ is continuous, so $f * g$ must vanish identically on $[0, 1]$. Hence

$$0 = (f * g)(1) = \int_0^1 f(x)g(1 - x)\, dx.$$

This says that the function $h(x) = \overline{g(1 - x)}$ is orthogonal to f. Moreover, if n is any positive integer and χ_n denotes the convolution of χ with itself $n - 1$ times, then

$$\sigma(\chi_n * f * g) \geqslant \sigma(\chi_n) + \sigma(f * g) = \sigma(f * g) \geqslant 1,$$

so that the above reasoning shows h is orthogonal to the projection onto $L^2(0, 1)$ of $\chi_n * f$. The latter function is $V^n f$, and thus h is orthogonal to the invariant subspace of V generated by f. That subspace is one of the subspaces $L^2(a, 1)$, $0 \leqslant a \leqslant 1$. But since $\sigma(f) = 0$ it must in fact be $L^2(0, 1)$, so that $h = 0$. Therefore $g = 0$, and the lemma is established.

As for the theorem, one can easily reduce the general case to the case where $\sigma(f) = 0$, $\sigma(g) \geqslant 0$, and $\sigma(f * g) > 0$. Assuming these conditions are fulfilled, let b denote $\sigma(f * g)$ if the latter is finite and an arbitrary positive real number otherwise. We must show that $\sigma(g) \geqslant b$.

Let k be the characteristic function of $(0, \infty)$, and let $f_1 = k * f$ and $g_1 = k * g$. The functions f_1 and g_1 are locally absolutely continuous and their derivatives are (almost everywhere) f and g; thus $\sigma(f_1) = \sigma(f) = 0$ and $\sigma(g_1) = \sigma(g)$. We note that

$$\sigma(f_1 * g_1) \geqslant \sigma(f * g) + \sigma(k * k) = \sigma(f * g) \geqslant b. \tag{16}$$

Let f_2 be the function that equals f_1 on $(0, b)$ and equals 0 elsewhere, and let g_2 be defined analogously relative to g_1. Then $f_1 * g_1$ and $f_2 * g_2$ coincide in $(0, b)$; so $f_2 * g_2$ vanishes there by (16). The functions f_2 and g_2 belong to $L^2(0, b)$, so we can apply the lemma to conclude that $g_2 = 0$. Hence $\sigma(g_1) \geqslant b$, and thus $\sigma(g) \geqslant b$ also. The theorem is proved.

The reader will find references to other proofs of Titchmarsh's theorem in Kalisch's paper. A weakened version of the theorem of Titchmarsh plays a crucial role in Mikusiński's operational calculus, and Mikusiński's book [27] contains a real variables proof of this weakened version due to Ryll-Nardzewski.

To conclude this section I wish to mention an interesting observation on the Volterra operator due to Radjavi and Rosenthal. Let X be the operator on $L^2(0, 1)$ of multiplication by x, and let $\mathcal{A}$ be the weakly closed operator algebra generated by X and V. Since $\operatorname{Lat} V \subset \operatorname{Lat} X$, we have $\operatorname{Lat} \mathcal{A} = \operatorname{Lat} V$; thus $\operatorname{Lat} \mathcal{A}$ is linearly ordered. Moreover $\mathcal{A}$ contains a maximal commutative selfadjoint operator algebra, namely, the weakly closed operator algebra generated by X. (The maximality of this algebra follows from Theorem 1.) The theorem of Radjavi and Rosenthal mentioned near the end of §4 can thus be applied to give the following conclusion: *If A is an operator on $L^2(0, 1)$ which leaves invariant all the subspaces $L^2(a, 1), 0 \leqslant a \leqslant 1$, then A is contained in $\mathcal{A}$.*

9. The Volterra operator plus multiplication by x. In this concluding section we consider another of the handful of operators whose invariant subspaces can be completely described, the operator $X + V$. Here, as before, V stands for the Volterra operator on $L^2(0, 1)$ and X stands for the operator on $L^2(0, 1)$ of multiplication by x. I do not know of any place in the literature where the invariant subspaces of $X + V$ are treated, even though, as we shall see, the discussion is very elementary.

In practice it is more convenient to work with a slightly different operator. Let m be the sum of Lebesgue measure on $(0, 1)$ and a unit point mass at 0. Let the operator T on $L^2(m)$ be defined by

$$(Tf)(x) = \begin{cases} 0 & \text{for } x = 0, \\ xf(x) + \int_0^x f(t)\, dm(t) & \text{for } x > 0. \end{cases}$$

The space $L^2(0, 1)$, when regarded in the usual way as a subspace of $L^2(m)$, is invariant under T, and the restriction of T to $L^2(0, 1)$ is $X + V$. Hence, the problem of finding the invariant subspaces of $X + V$ is contained in the problem of finding the invariant subspaces of T.

A simple observation enables us to replace the latter problem with the problem of finding the closed ideals in a certain Banach algebra. Let V_1 be the operator on $L^2(m)$ analogous to V; that is,

$$(V_1 f)(x) = \int_0^x f(t)\, dm(t) = f(0) + \int_0^x f(t)\, dt.$$

The range of V_1, which we denote by R, consists of all absolutely continuous functions on $[0, 1]$ whose derivatives belong to $L^2(0, 1)$, and V_1 provides a one-to-one correspondence between $L^2(m)$ and R. The inverse of V_1 is given by

$$(V_1^{-1} g)(x) = \begin{cases} g(x) & \text{for } x = 0, \\ g'(x) & \text{for } x > 0. \end{cases}$$

Now suppose f is in $L^2(m)$, and let $g = V_1 f$. Then for $x > 0$ we have

$$\begin{aligned} (Tf)(x) &= xf(x) + (V_1 f)(x) \\ &= xg'(x) + g(x) \\ &= \frac{d}{dx}\,[xg(x)]\,. \end{aligned}$$

In other words, if we let X_1 denote the operator on R of multiplication by x, then $(Tf)(x) = (V_1^{-1} X_1 V_1 f)(x), x > 0$. The equality holds also at $x = 0$, because both sides vanish there. Thus, V_1 transforms the operator T into the operator on R of multiplication by x.

We norm R by declaring that the map V_1 shall be an isometry, in other words, by putting $\|g\|$ equal to the norm of $V_1^{-1} g$ in $L^2(m)$. This makes R into a Hilbert space. We note that if g is in R and $f = V_1^{-1} g$, then

$$|g(x)| \leqslant \int_0^x |f(t)|\, dm(t)$$

$$(17) \qquad \leqslant \left(\int_0^1 |f(t)|^2\, dm(t)\right)^{1/2} \left(\int_0^1 1^2\, dm(t)\right)^{1/2}$$

$$= 2^{1/2} \|g\|, \quad 0 \leqslant x \leqslant 1.$$

In particular, the evaluation functionals on R at points of $[0, 1]$ are continuous.

The space R is not merely a linear space, it is also an algebra; that is, it is closed under multiplication. If g and h are in R then, using (17), we obtain

$$\|gh\|^2 = |g(0)h(0)|^2 + \int_0^1 |(gh)'|^2\, dx$$

$$\leqslant |g(0)h(0)|^2 + 2\int_0^1 |g|^2|h'|^2\, dx + 2\int_0^1 |g'|^2|h|^2\, dx$$

$$\leqslant \|g\|^2\|h\|^2 + 4\|g\|^2\int_0^1 |h'|^2\, dx + 4\|h\|^2\int_0^1 |g'|^2\, dx$$

$$\leqslant 9\|g\|^2\|h\|^2;$$

in other words $\|gh\| \leqslant 3\|g\|\,\|h\|$. This inequality shows that multiplication in R is norm continuous. The inequality $\|gh\| \leqslant \|g\|\,\|h\|$ does not hold in general, as one can show with simple examples; thus R is not quite a Banach algebra. We could make R into a Banach algebra by introducing an equivalent norm, but that will not be necessary for our purposes.

A subspace S of $L^2(m)$ is invariant under T if and only if V_1S is invariant under multiplication by x. If the latter happens then V_1S is invariant under multiplication by all polynomials. But the polynomials are dense in R, because, obviously, their images under V_1^{-1} are dense in $L^2(m)$. Thus, the image under V_1 of an invariant subspace of T is a closed ideal in R. The converse of the last statement is trivial, so the problem of finding the invariant subspaces of T is equivalent to the problem of finding the closed ideals in R. (Incidentally, my purpose in introducing the operator T was to enable us to work in an algebra with an identity. One can carry out the above manipulations working with $X + V$ instead of T, but one must then replace R by the subalgebra of functions in R that vanish at 0.)

Before trying to determine the most general closed ideal in R we shall find the maximal ideals, that is, the ideals of codimension one. These, it turns out, are in one-to-one correspondence with the points of $[0, 1]$, each point x

corresponding to the ideal of functions in R that vanish at x. Although this result follows immediately from a general lemma about Banach function algebras [25, p. 55], the following alternative proof might be instructive.

A maximal ideal in R is the image under V_1 of an invariant subspace of T of codimension one. The orthogonal complement of such a subspace is an invariant subspace of T^* of dimension one, so it is spanned by an eigenvector of T^*. Our task, therefore, is reduced to finding the eigenvectors of T^*.

A straightforward calculation produces the formula

$$\begin{aligned}(T^*f)(x) &= \int_0^1 f(t)\,dt && \text{for } x = 0,\\ &= xf(x) + \int_x^1 f(t)\,dt && \text{for } x > 0.\end{aligned}$$

Hence the equation $T^*f = \lambda f$ (λ a complex number) is equivalent to the equation

$$(\lambda - x)f(x) = \int_x^1 f(t)\,dt, \qquad 0 \leqslant x \leqslant 1. \tag{18}$$

If λ is not in the interval $[0, 1]$, then (18) implies that f is absolutely continuous on $[0, 1]$. Moreover, differentiation of (18) in this case shows that $f' = 0$ almost everywhere, so that f must vanish identically. All eigenvalues of T^*, therefore, lie in the interval $[0, 1]$.

If λ is in $[0, 1]$ then the preceding reasoning shows that f must be absolutely continuous with a vanishing derivative on each of the intervals $[0, \lambda)$ and $(\lambda, 1]$, so that f is constant on each of these intervals. (This statement must be modified in the obvious ways for the extreme values $\lambda = 0, 1$.) Moreover, if $\lambda \neq 1$ then (18) implies that $f(1) = 0$. Thus, for a fixed λ in $[0, 1]$, any eigenvector of T^* for the eigenvalue λ must be a constant multiple of the function

$$\begin{aligned}\varphi_\lambda(x) &= 1 && \text{for } 0 \leqslant x \leqslant \lambda,\\ &= 0 && \text{for } \lambda < x \leqslant 1.\end{aligned}$$

It is obvious that φ_λ is actually a solution of (18), and hence each λ in $[0, 1]$ is an eigenvalue of T^* of unit multiplicity. The orthogonal complement of φ_λ is mapped by V_1 onto the set of functions in R that vanish at λ; this verifies the description given above of the maximal ideals in R.

The closed ideals in R, we shall show, are in one-to-one correspondence with the closed subsets of $[0, 1]$, each closed set corresponding to the ideal of functions in R that vanish on it. In order to state this result in a more convenient form, we recall that if I is an ideal in R, then the *hull* of I, denoted $\text{hull}(I)$, is by definition the (obviously closed) set of all points in $[0, 1]$ at which every function in I vanishes. If K is a closed subset of $[0, 1]$, then the *kernel* of K, denoted $\ker(K)$, is by definition the ideal of all functions in R that vanish everywhere on K. It is clear that distinct closed subsets of $[0, 1]$ have distinct kernels.

The following theorem restates the above description of the closed ideals in R.

THEOREM 7. *If I is a closed ideal in R, then I is the kernel of its hull.*

This theorem establishes a one-to-one correspondence between the invariant subspaces of T and the closed subsets of $[0, 1]$. The correspondence reverses inclusion. Hence the lattice $\text{Lat } T$ is isomorphic to the lattice of relatively open subsets of $[0, 1]$. The lattice $\text{Lat } T^*$ is isomorphic to the lattice of closed subsets of $[0, 1]$.

It is convenient to break up the proof of the theorem into two lemmas.

LEMMA 1. *Let I be an ideal in R and let g be a function in R whose support is disjoint from $\text{hull}(I)$. Then g belongs to I.*

LEMMA 2. *Let K be a closed subset of $[0, 1]$ and let h be a function in $\ker(K)$. Then h is the limit, in the norm of R, of a sequence of functions in R whose supports are disjoint from K.*

These lemmas together clearly yield the theorem. We begin with Lemma 1, which is actually an instance of a general result about regular function algebras [**25**, p. 84]. In our special case a simpler and more direct (although not essentially different) proof can be given.

Let I and g be as in Lemma 1, and let C denote the support of g. Since C does not meet $\text{hull}(I)$, there is, for each point of C, a function in I that is nonzero at the point. This together with the compactness of C enables us to select finitely many functions $h_1, h_2, \cdots, h_q$ in I which do not vanish simultaneously at any point of C. The function $u = |h_1|^2 + |h_2|^2 + \cdots + |h_q|^2$ then belongs to I and is nonzero at every point of C. Choose a nonnegative function v in R such that v vanishes on C but is nonzero at every point where u vanishes. (I leave it to the reader to convince himself that such a function exists. One can even take it to be of class C^∞.) The function $u + v$ belongs to R and is everywhere positive. By an elementary argument it follows that $(u + v)^{-1}$ is in R. Hence, the function $w = u(u + v)^{-1}$ is in I. Since

$w = 1$ everywhere on C, we may conclude that $g = gw$ belongs to I. This proves Lemma 1.

The proof of Lemma 2 is computational and basically trivial. Let K and h be as in the statement of that lemma. We shall assume that K contains neither 0 nor 1; the proof that follows must be modified in an obvious way if K contains one or both of the latter points. We show that, given $\epsilon > 0$, there is a function g in R whose support is disjoint from K such that $\|h - g\| < \epsilon$; this will establish the lemma. Let $f = V_1^{-1}h$. We shall define g to be $V_1 f_1$ where f_1 is a slight modification of f. We note that the relation $f = h'$, which holds almost everywhere in $(0, 1]$, implies that f vanishes almost everywhere on K.

Let $[0, b), (a, 1], (a_1, b_1), (a_2, b_2), \cdots$ be the complementary intervals of K in $[0, 1]$. Choose a positive integer N such that

$$\sum_{n>N} \int_{a_n}^{b_n} |f(x)|^2\, dx < \epsilon^2/4. \tag{19}$$

Next, choose closed subintervals $[0, \beta], [\alpha, 1], [\alpha_1, \beta_1], \cdots, [\alpha_N, \beta_N]$ of $[0, b), (a, 1], (a_1, b_1), \cdots, (a_N, b_N)$, respectively, with $\beta > b/2$, $1 - \alpha > (1 - a)/2$, $(\beta_n - \alpha_n) > (b_n - a_n)/2$ for each n, such that

$$\int_{\beta}^{b} |f(x)|^2\, dx < \epsilon^2/4(N + 2), \tag{20}$$

$$\int_{a}^{\alpha} |f(x)|^2\, dx < \epsilon^2/4(N + 2), \tag{21}$$

$$\int_{a_n}^{\alpha_n} |f(x)|^2\, dx + \int_{\beta_n}^{b_n} |f(x)|^2\, dx < \epsilon^2/4(N + 2), \qquad n = 1, \cdots, N. \tag{22}$$

Define the function f_1 in $L^2(m)$ to be 0 except as indicated below:

$$f_1(0) = h(0),$$

$$f_1(x) = f(x) - \frac{1}{\beta}\left[h(0) + \int_0^{\beta} f(y)\, dy\right], \qquad 0 < x < \beta,$$

$$f_1(x) = f(x) + \frac{1}{1 - \alpha}\left[h(1) - \int_{\alpha}^{1} f(y)\, dy\right], \qquad \alpha < x < 1,$$

$$f_1(x) = f(x) - \frac{1}{\beta_n - \alpha_n}\int_{\alpha_n}^{\beta_n} f(y)\, dy, \qquad \alpha_n < x < \beta_n,\ n = 1, \cdots, N.$$

The support of $g = V_1 f_1$ is then contained in the union of the intervals $[0, \beta]$, $[\alpha, 1]$, $[\alpha_1, \beta_1], \cdots, [\alpha_N, \beta_N]$; thus, this support is disjoint from K. To complete the proof it remains to show that $\|h - g\| < \epsilon$, in other words, that the norm of $f - f_1$ in $L^2(m)$ is less than ϵ. The square of the norm of $f - f_1$ can be written as the sum of two terms, M_1 and M_2, where M_1 is the sum of the quantities on the left sides of the inequalities (19)–(22), and

$$M_2 = \frac{1}{\beta}\left|h(0) + \int_0^\beta f(x)\,dx\right|^2 + \frac{1}{1-\alpha}\left|h(1) - \int_\alpha^1 f(x)\,dx\right|^2 + \sum_{n=1}^{N} \frac{1}{\beta_n - \alpha_n}\left|\int_{\alpha_n}^{\beta_n} f(x)\,dx\right|^2.$$

From (19)–(22) we have $M_1 < \epsilon^2/2$. We complete the proof by showing that $M_2 < \epsilon^2/2$.

The vanishing of h on K implies the equalities

$$h(0) + \int_0^b f(x)\,dx = 0,$$

$$\int_a^1 f(x)\,dx = h(1),$$

$$\int_{a_n}^{b_n} f(x)\,dx = 0, \qquad n = 1, \cdots, N.$$

Using these, we can rewrite M_2 as

$$M_2 = \frac{1}{\beta}\left|\int_\beta^b f(x)\,dx\right|^2 + \frac{1}{1-\alpha}\left|\int_a^\alpha f(x)\,dx\right|^2 + \sum_{n=1}^{N} \frac{1}{\beta_n - \alpha_n}\left|\int_{a_n}^{\alpha_n} f(x)\,dx + \int_{\beta_n}^{b_n} f(x)\,dx\right|^2.$$

By a simple application of Schwarz's inequality,

$$\left|\int_\beta^b f(x)\,dx\right|^2 \leqslant b \int_\beta^b |f(x)|^2\,dx,$$

$$\left|\int_a^\alpha f(x)\,dx\right|^2 \leqslant (1-a) \int_a^\alpha |f(x)|^2\,dx,$$

$$\left|\int_{a_n}^{\alpha_n} f(x)\,dx + \int_{\beta_n}^{b_n} f(x)\,dx\right|^2 \leqslant (b_n - a_n)\left[\int_{a_n}^{\alpha_n} |f(x)|^2\,dx + \int_{\beta_n}^{b_n} |f(x)|^2\,dx\right].$$

These together with (20)–(22) and the inequalities $b < 2\beta$, $1 - a < 2(1 - \alpha)$, $b_n - a_n < 2(\beta_n - \alpha_n)$, give

$$M_2 < \frac{\epsilon^2}{2(N+2)} + \frac{\epsilon^2}{2(N+2)} + \frac{N\epsilon^2}{2(N+2)} = \frac{\epsilon^2}{2}.$$

The proof of the lemma is now complete.

BIBLIOGRAPHY

1. N. Aronszajn and K. T. Smith, *Invariant subspaces of completely continuous operators*, Ann. of Math. (2) **60** (1954), 345–350. MR **16**, 488.

2. W. B. Arveson, *A density theorem for operator algebras*, Duke Math. J. **34** (1967), 635–647. MR **36** #4345.

3. W. B. Arveson and J. Feldman, *A note on invariant subspaces*, Michigan Math. J. **15** (1968), 61–64. MR **36** #6969.

4. A. Beurling, *On two problems concerning linear transformations in Hilbert space*, Acta Math. **81** (1949), 239–255.

5. L. Brickman and P. A. Fillmore, *The invariant subspace lattice of a linear transformation*, Canad. J. Math. **19** (1967), 810–822. MR **35** #4242.

6. M. S. Brodskiĭ, *On a problem of I. M. Gel′fand*, Uspehi Mat. Nauk **12** (1957), no. 2 (74), 129–132. (Russian) MR **20** #1229.

7. C. Davis, H. Radjavi and P. Rosenthal, *On operator algebras and invariant subspaces*, Canad. J. Math. **21** (1969), 1178–1181. MR **40** #7847.

8. J. Dixmier, *Les algèbres d'opérateurs dans l'espace Hilbertien* (*Algèbres de von Neumann*), Cahiers scientifiques, fasc. 25, Gauthier-Villars, Paris, 1957. MR **20** #1234.

9. W. F. Donoghue, Jr., *The lattice of invariant subspaces of a completely continuous quasi-nilpotent transformation*, Pacific J. Math. **7** (1957), 1031–1036. MR **19**, 1066.

10. N. Dunford and J. T. Schwartz, *Linear operators*. Part II: *Spectral theory. Selfadjoint operators in Hilbert space*, Interscience, New York, 1963. MR **32** #6181.

11. T. W. Gamelin, *Uniform algebras*, Prentice-Hall, Englewood Cliffs, N. J., 1962.

12. R. Goodman, *Invariant subspaces for normal operators*, J. Math. Mech. **15** (1966), 123–128. MR **32** #4543.

13. P. R. Halmos, *Measure theory*, Van Nostrand, Princeton, N. J., 1950. MR **11**, 504.

14. ———, *Shifts on Hilbert spaces*, J. Reine Angew. Math. **208** (1961), 102–112. MR **27** #2868.

15. ———, *Invariant subspaces of polynomially compact operators*, Pacific J. Math. **16** (1966), 433–437. MR **33** #1725.

16. ———, *A Hilbert space problem book*, Van Nostrand, Princeton, N. J., 1967. MR **34** #8178.

17. H. Helson, *Lectures on invariant subspaces*, Academic Press, New York, 1964. MR **30** #1409.

18. H. Helson and D. Lowdenslager, *Prediction theory and Fourier series in several variables*, Acta Math. **99** (1958), 165–202. MR **20** #4155.

19. ———, *Invariant subspaces*, Proc. Internat. Sympos. Linear Spaces (Jerusalem, 1960), Jerusalem Academic Press, Jerusalem; Pergamon, Oxford, 1961, pp. 251–265. MR **28** #487.

20. E. Hille and R. S. Phillips, *Functional analysis and semi-groups*, rev. ed., Amer. Math. Soc. Colloq. Publ., vol. 31, Amer. Math. Soc., Providence, R. I., 1957. MR **19**, 664.

21. K. Hoffman, *Banach spaces of analytic functions*, Prentice-Hall Ser. in Modern Analysis, Prentice-Hall, Englewood Cliffs, N. J., 1962. MR **24** #A2844.

22. K. Hoffman and I. M. Singer, *Maximal algebras of continuous functions*, Acta Math. **103** (1960), 217–241. MR **22** #8318.

23. G. K. Kalisch, *A functional analysis proof of Titchmarsh's theorem on convolution*, J. Math. Anal. Appl. **5** (1962), 176–183. MR **25** #4307.

24. P. Lax, *Translation invariant spaces*, Acta Math. **101** (1959), 163–178. MR **21** #4359.

25. L. H. Loomis, *An introduction to abstract harmonic analysis*, Van Nostrand, Princeton, N. J., 1953. MR **14**, 883.

26. G. W. Mackey, *A theorem of Stone and von Neumann*, Duke Math. J. **16** (1949), 313–326. MR **11**, 10.

27. Ja. G. Mikusiński, *The calculus of operators*, Monografie Mat., vol. 30, PWN, Warsaw, 1953; English transl., Interscience Ser. of Monographs on Pure and Appl. Math., vol. 8, Pergamon Press, New York; PWN, Warsaw, 1959. MR **16**, 243; **21** #4333.

28. E. A. Nordgren, *Transitive operator algebras*, J. Math. Anal. Appl. **32** (1970), 639–643. MR **42** #6627.

29. E. A. Nordgren, H. Radjavi and P. Rosenthal, *On density of transitive algebras*, Acta Sci. Math. (Szeged) **30** (1969), 175–179. MR **40** #6276.

30. ———, *On operators with reducing invariant subspaces*, Amer. J. Math. (to appear)

31. H. Radjavi and P. Rosenthal, *On invariant subspaces and reflexive algebras*, Amer. J. Math. **91** (1969), 683–692. MR **40** #4796.

32. F. Riesz and M. Riesz, *Über die Randwerte einer analytischen Funktion*, IV Skand. Mat. Kongr. (Stockholm, 1916), Mittag-Leffler, Uppsala, 1920, pp. 27–44.

33. M. Rosenblum, *On a theorem of Fuglede and Putnam*, J. London Math. Soc. **33** (1958), 376–377. MR **20** #6037.

34. P. Rosenthal, *Examples of invariant subspace lattices*, Duke Math. J. **37** (1970), 103–112. MR **40** #4797.

35. W. Rudin, *Real and complex analysis*, McGraw-Hill, New York, 1966. MR **35** #1420.

36. D. Sarason, *A remark on the Volterra operator*, J. Math. Anal. Appl. **12** (1965), 244–246. MR **33** #580.

37. ———, *Invariant subspaces and unstarred operator algebras*, Pacific J. Math. **17** (1966), 511–517. MR **33** #590.

38. ———, *Representing measures for* $R(X)$ *and their Green's functions*, J. Functional Anal. 7 (1971), 359–385.

39. I. E. Segal, *Decompositions of operator algebras*. II. *Multiplicity theory*, Mem. Amer. Math. Soc. No. 9 (1951). MR **13**, 472.

40. J. T. Schwartz, *Subdiagonalization of operators in Hilbert space with compact imaginary part*, Comm. Pure Appl. Math. **15** (1962), 159–172. MR **26** #1759.

41. J. Wermer, *The existence of invariant subspaces*, Duke Math. J. **19** (1952), 615–622. MR **14**, 384.

42. ———, *On algebras of continuous functions*, Proc. Amer. Math. Soc. **4** (1953), 866–869. MR **15**, 440.

UNIVERSITY OF CALIFORNIA, BERKELEY

Mathematical Surveys
Volume 13
1974

II

WEIGHTED SHIFT OPERATORS AND ANALYTIC FUNCTION THEORY

BY

ALLEN L. SHIELDS

AMS (MOS) subject classifications (1970). Primary 47B99; Secondary 47C05, 46L15.

ALLEN L. SHIELDS

1. Introduction. A *weighted shift* operator T on (complex) Hilbert space H is an operator that maps each vector in some orthonormal basis $\{e_n\}$ into a scalar multiple of the next vector

$$Te_n = w_n e_{n+1} \tag{1}$$

for all n. If the index n runs over the nonnegative integers, then T is called a *unilateral weighted shift*, while if n runs over all integers, then T is called a *bilateral weighted shift*. We shall treat both cases. In the main part of the article we shall not try to assign credits for most of the results. At the end we shall say something about the history of the subject.

The reader is assumed to be familiar with the elementary facts about operators on Hilbert space. Also, some basic facts from the theory of commutative Banach algebras with identity are used, as are some facts about analytic functions of one complex variable.

The study was motivated by the theory of the unweighted unilateral and bilateral shifts ($w_n = 1$, all n). Some familiarity with this theory is desirable, though not essential. (See, for example, §6 of the article by Sarason in this volume, or Chapters I–III in the book by Helson **[39]**, or Chapter 7 in the book by Hoffman **[46]**.)

In what follows T will always denote a weighted shift operator with weight sequence $\{w_n\}$. We shall sometimes omit the adjective "weighted" and refer to T simply as a shift.

A number of unsolved problems are stated as *Questions* scattered through the text.

2. Elementary properties.

PROPOSITION 1. *If* $\{\lambda_n\}$ *are complex numbers of modulus* 1, *then* T *is unitarily equivalent to the weighted shift operator with weight sequence* $\{\bar{\lambda}_{n+1}\lambda_n w_n\}$.

PROOF. Let U be the unitary operator defined by $Ue_n = \lambda_n e_n$. Then the operator U^*TU is a weighted shift with the weight sequence indicated above.

COROLLARY 1. *T is unitarily equivalent to the weighted shift operator with weight sequence* $\{|w_n|\}$.

PROOF. Choose $\lambda_0 = 1$; the choice of the remaining λ_n is then forced (in both the unilateral and the bilateral cases).

COROLLARY 2. *If* $|c| = 1$, *then T and cT are unitarily equivalent.*

PROOF. Let $\lambda_n = \bar{c}^n$, for all n, in Proposition 1.

From Corollary 2 we see that the spectrum of T, as well as the various parts of the spectrum, have circular symmetry about the origin.

PROPOSITION 2. $\|T^n\| = \sup_k |w_k w_{k+1} \cdots w_{k+n-1}|$, $n = 1, 2, \cdots$.

PROOF. From (1) we have

$$T^n e_k = (w_k w_{k+1} \cdots w_{k+n-1}) e_{k+n} \qquad (\text{all } k), \tag{2}$$

and the result follows.

In particular the linear transformation T, defined on finite linear combinations of basis vectors by (1), is bounded if and only if the weight sequence is bounded.

The following proposition is trivial and we omit its proof.

PROPOSITION 3. *If T is a bilateral weighted shift then*

$$T^* e_n = \bar{w}_{n-1} e_{n-1} \qquad (\text{all } n).$$

If T is a unilateral weighted shift then

$$T^* e_n = \bar{w}_{n-1} e_{n-1} \qquad (n \geqslant 1),$$

$$T^* e_0 = 0.$$

A compact operator A on Hilbert space is said to be in the class C^p $(0 < p < \infty)$ if the eigenvalues of $(A^*A)^{1/2}$ are in l^p.

PROPOSITION 4. *T is compact if and only if* $w_n \to 0$ $(|n| \to \infty)$; $T \in C^p$ *iff* $\{w_n\} \in l^p$.

PROOF. A calculation shows that

$$(T^*T)^{1/2} e_n = |w_n| e_n \tag{3}$$

for all n. The result for C^p is immediate, and the compactness criterion follows since T is compact if and only if $(T^*T)^{1/2}$ is compact, and a diagonal operator is compact if and only if the diagonal entries tend to zero. (Alternatively, the compactness criterion can be established directly without invoking (3).)

In the next proposition S and T are weighted shifts with respect to the same orthonormal basis $\{e_n\}$, and with weight sequences $\{v_n\}$ and $\{w_n\}$ respectively. Also, A denotes an operator on H and (a_{ij}) denotes its matrix with respect to the basis $\{e_n\}$: $a_{ij} = (Ae_j, e_i)$.

In §4 we shall study in detail the operators that commute with a weighted shift. In the following proposition we consider a more general situation.

PROPOSITION 5. (a) *If S and T are bilateral shifts, then $AS = TA$ if and only if*

$$v_j a_{i+1,j+1} = w_i a_{ij} \qquad (\text{all } i, j). \tag{4}$$

(b) *If S and T are unilateral shifts, then $AS = TA$ if and only if*

$$\begin{aligned} v_j a_{i+1,j+1} &= w_i a_{ij} \qquad (\text{all } i, j \geqslant 0), \\ v_j a_{0,j+1} &= 0 \qquad (\text{all } j \geqslant 0). \end{aligned} \tag{5}$$

PROOF. (a) For all i, j we have

$$(ASe_j, e_{i+1}) = v_j(Ae_{j+1}, e_{i+1}) = v_j a_{i+1,j+1},$$

$$(TAe_j, e_{i+1}) = (Ae_j, T^*e_{i+1}) = w_i a_{ij},$$

and the result follows.

(b) The proof is similar and we omit it.

Using this we can discuss similarity and unitary equivalence of weighted shifts. We use the notations of the preceding proposition.

THEOREM 1. (a) *If S, T are two bilateral weighted shifts with weight sequences $\{v_n\}$, $\{w_n\}$, and if there exists an integer k such that*

$$|v_n| = |w_{n+k}|, \quad \text{for all } n, \tag{6}$$

then S and T are unitarily equivalent. The converse is valid provided that none of the weights is zero.

(b) *If S and T are unilateral shifts and if $|v_n| = |w_n|$ for all n, then S and T are unitarily equivalent. The converse is valid if none of the weights is zero.*

PROOF. (a) If (6) holds, then we define A by $Ae_n = \alpha_n e_{n+k}$ for all n, where $\{\alpha_n\}$ are defined inductively from the formulae:

$$\alpha_0 = 1, \qquad v_n \alpha_{n+1} = \alpha_n w_{n+k} \qquad (n = 0, \pm 1, \pm 2, \cdots).$$

(When zero weights occur, the corresponding α_n is defined to be 1.) Then $|\alpha_n| = 1$ for all n and so A is unitary; one verifies that $AS = TA$.

Conversely, suppose that none of the weights is zero and that $AS = TA$ for some unitary operator A. Then we have $A^*T = SA^*$. Let (a_{ij}) and (b_{ij}) denote the matrices of A and A^* with respect to the basis (e_n) (thus $a_{ij} = \overline{b}_{ji}$). From Proposition 5 we have

$$v_j a_{i+1,j+1} = w_i a_{ij}, \qquad w_m b_{n+1,m+1} = v_n b_{nm},$$

for all i, j, n, m. Hence

$$(w_i/v_j)a_{ij} = a_{i+1,j+1} = \overline{b}_{j+1,i+1} = (\overline{v}_j/\overline{w}_i)\overline{b}_{ji} = (\overline{v}_j/\overline{w}_i)a_{ij}.$$

Thus $|w_i| = |v_j|$ for all i, j such that $a_{ij} \neq 0$. Since $Ae_0 \neq 0$, there is an integer k such that $a_{k0} \neq 0$. From (4) we see that $a_{k+n,n} \neq 0$ for all n, and this implies (6).

(b) The proof is similar and we omit it (see the discussion preceding Problem 76 in [**34**]).

THEOREM 2. (a) *Let S, T be bilateral weighted shifts with weight sequences $\{v_n\}$, $\{w_n\}$, and with no zero weights. Then S and T are similar if and only if there exists an integer k and positive constants C_1 and C_2 such that*

(7) $$0 < C_1 \leqslant |w_{k+m} \cdots w_{k+n}/v_m \cdots v_n| \leqslant C_2 \qquad (\text{all } m \leqslant n).$$

(b) *Let S, T be unilateral shifts with no zero weights. Then S and T are similar if and only if (7) holds with $k = m = 0$, for all $n \geqslant 0$.*

In both of these cases the operator A that implements the similarity can be chosen so that

$$\max(\|A\|, \|A^{-1}\|) \leqslant \max(C_2, 1/C_1).$$

PROOF. (a) Assume that (7) holds, and let $\beta_n = w_{k+n}/v_n$. Let $\alpha_0 = 1$ and

$$\alpha_n = \beta_0\beta_1 \cdots \beta_{n-1} \qquad (n > 0),$$

$$\alpha_n = (\beta_{-1}\beta_{-2} \cdots \beta_n)^{-1} \qquad (n < 0).$$

From (7) it follows that the sequences $\{\alpha_n\}$ and $\{1/\alpha_n\}$ are both bounded. Define A by $Ae_n = \alpha_n e_{n+k}$ (all n). Then A is an invertible operator and $AS = TA$. Further, A and A^{-1} are bounded by the maximum of C_2 and $1/C_1$.

Conversely, let A be an invertible operator such that $AS = TA$. Then also $A^{-1}T = SA^{-1}$. Let the matrices for A and A^{-1} be (a_{ij}) and (b_{ij}) respectively.

From (4) we have

$$a_{k+n,n} = \beta_{n-1} a_{k+n-1,n-1} \qquad \text{(all } n). \tag{8}$$

From this we obtain

$$\begin{cases} a_{k+n,n} = \beta_{n-1} \cdots \beta_0 a_{k0} & (n > 0), \\ a_{k+n-1,n-1} = (\beta_{n-1}\beta_n \cdots \beta_{-1})^{-1} a_{k0} & (n < 0). \end{cases} \tag{9}$$

Applying (4) to $A^{-1}T = SA^{-1}$ we have $b_{n,k+n} = (1/\beta_{n-1})b_{n-1,k+n-1}$. From this we obtain

$$\begin{cases} b_{n,k+n} = (\beta_{n-1} \cdots \beta_0)^{-1} b_{0k} & (n > 0), \\ b_{n-1,k+n-1} = \beta_{n-1}\beta_n \cdots \beta_{-1} b_{0k} & (n < 0). \end{cases} \tag{10}$$

Next observe that $\Sigma\, b_{0k}a_{k0} = 1$ (since $A^{-1}A = I$) and so there exists an integer k such that $a_{k0} \neq 0$ and $b_{0k} \neq 0$. Fix this k. Finally, $|a_{ij}| \leqslant \|A\|$, $|b_{ij}| \leqslant \|A^{-1}\|$ for all i, j. Applying this to (9) and (10) we obtain (7), at least when either m or n is zero. The general case follows from this.

(b) The proof is similar and we omit it (see Problem 76 in **[34]**).

DEFINITION. An operator A is said to be *power bounded* if $\|T^n\| \leqslant c$, $n = 1, 2, \cdots$.

COROLLARY. *If T is a weighted shift (or an arbitrary direct sum of weighted shifts) and if T is power bounded, then T is similar to a contraction.*

PROOF. The hypothesis implies that there is a constant c such that $|w_k \cdots w_{k+n}| \leqslant c$ (for all k, n) where $\{w_n\}$ is the weight sequence for T, if T itself is a weighted shift, otherwise $\{w_n\}$ is the weight sequence of any of the weighted shifts that are direct summands of T.

We first assume that T is a unilateral shift with all weights different from zero. Let $\pi(n) = w_0 \cdots w_n$ $(n \geqslant 0)$. We shall define a new weight sequence $\{v(n)\}$ such that the unilateral shift S with these weights is similar to T. If $\pi(n) \geqslant 1$ for all n, then we let $v(n) = 1$ for all n; in this case it is easy to check that the similarity criterion of Theorem 2(b) is fulfilled.

Otherwise, let n_1 be the first integer such that $\pi(n_1) < 1$. Let $v(k) = 1$ for all $k < n_1$ and let $v(n_1) = \pi(n_1)$. If $\pi(n)/\pi(n_1) \geqslant 1$ for all $n > n_1$, then we let $v(n) = 1$ for all such n. In this case it is easy to check that S and T are similar.

Otherwise, let n_2 be the first integer greater that n_1 such that $\pi(n_2)/\pi(n_1) < 1$. Let $v(k) = 1$ for $n_1 < k < n_2$ (if there are any such k). Let $v(n_2) = \pi(n_2)/\pi(n_1)$.

Continue choosing the weights $v(n)$ in this manner. The operator S so defined will be a contraction, i.e., $\|S\| \leqslant 1$, since all $v_n \leqslant 1$. Also, S will be similar to T.

The case when T is a bilateral weighted shift with no zero weights can be handled in an analogous manner. Furthermore, in both of these cases if A is the operator that implements the similarity then by Theorem 2 we may bound $\|A\|$ and $\|A^{-1}\|$ in terms of c. This allows us to handle any direct sum of weighted shifts (unilateral or bilateral) with no zero weights. Finally, the case of zero weights leads to direct sums involving finite-dimensional weighted shifts (see the discussion following this proof). A proof similar to the above shows that power bounded finite-dimensional weighted shifts are similar to contractions, and gives a bound on A and A^{-1}. We omit the details.

In Theorem 2 we assumed that none of the weights is zero. In a moment we shall limit ourselves to this case, but first we say a few words about shifts with some weights equal to zero. We make the following definition. An operator A on an n-dimensional Hilbert space $(n < \infty)$ is called a *finite-dimensional weighted shift* if there are numbers $\{w_1, \cdots, w_{n-1}\}$ and an orthonormal basis $\{e_1, \cdots, e_n\}$ such that

$$Ae_k = w_k e_{k+1} \qquad (k < n),$$

$$Ae_n = 0.$$

Such an operator is nilpotent (of degree n if no w_k is zero).

If T is a unilateral weighted shift and if one weight, say w_3, is zero, then T is the direct sum of a four-dimensional weighted shift (on the subspace spanned by $e_0, \cdots, e_3$) and an infinite-dimensional weighted shift (on the orthogonal complement). If finitely many weights are zero, then T is a direct sum of a finite number of finite-dimensional weighted shifts, plus one infinite-dimensional weighted shift. If infinitely many weights are zero, then T is the direct sum of an infinite family of finite-dimensional weighted shifts. This situation can be used to give an example of an operator with spectral radius 1, which is the direct sum of a countable family of operators (finite-dimensional nilpotent operators) each of which has spectrum $\{0\}$ (see [**34**, Problem 81]).

If T is a bilateral shift and if one weight, say w_0, is zero, then T is the direct sum of a unilateral weighted shift (on the subspace spanned by $\{e_k\}$, $k > 0$) and the adjoint of a unilateral weighted shift (on the orthogonal complement). If additional weights are zero, there is a further direct sum decomposition just as in the unilateral case.

Note that a weighted shift operator (unilateral or bilateral) is *injective* (i.e., one-to-one) if and only if none of the weights is zero.

From now on unless otherwise specified we shall assume that none of the weights is zero. By Corollary 1 to Proposition 1, in studying a single shift operator we may assume that the weights are nonnegative.

Assumption. $w_n > 0$ (all n).

3. **Weighted sequence spaces.** There is another way of viewing injective weighted shift operators amd we turn to this now. An *orthogonal basis* is a sequence $\{f_n\}$ of pairwise orthogonal vectors, none of them zero, which spans the space. An operator T is said to shift this orthogonal basis if $Tf_n = f_{n+1}$ (all n).

PROPOSITION 6. *T is an injective weighted shift operator if and only if T shifts some orthogonal basis.*

PROOF. Let T be a weighted shift operator with weight sequence $\{w_n\}$ with respect to the orthonormal basis $\{e_n\}$. Then T shifts the orthogonal basis $\{f_n\}$, where $f_0 = e_0$ and

(11) $$f_n = (w_0 w_1 \cdots w_{n-1}) e_n \qquad (n > 0)$$

and, in the case of a bilateral weighted shift,

(12) $$f_n = e_n / (w_n w_{n+1} \cdots w_{-1}) \qquad (n < 0)$$

(recall that we are assuming $w_n > 0$, all n).

Conversely, if T shifts the orthogonal basis $\{f_n\}$ then $Te_n = w_n e_{n+1}$, where

(13) $$e_n = f_n / \|f_n\|, \qquad w_n = \|f_{n+1}\| / \|f_n\| \qquad (\text{all } n).$$

This result will allow us to represent a weighted shift operator as an ordinary shift operator (that is, as "multiplication by z") on a Hilbert space of formal power series (in the unilateral case) or formal Laurent series (in the bilateral case). We now describe this point of view.

Let $\{\beta(n)\}$ be a sequence of positive numbers with $\beta(0) = 1$. We consider the space of sequences $f = \{\hat{f}(n)\}$ such that

(14) $$\|f\|^2 = \|f\|_\beta^2 = \sum |\hat{f}(n)|^2 [\beta(n)]^2 < \infty.$$

We shall use the notation

(15) $$f(z) = \sum \hat{f}(n)z^n$$

whether or not the series converges for any (complex) values of z. If n ranges only over the nonnegative integers these are formal power series; otherwise they are formal Laurent series. We shall denote these spaces as follows.

Notation. $H^2(\beta)$ (power series case).

$L^2(\beta)$ (Laurent series case).

These notations are suggested by the notations in the classical case $\beta \equiv 1$. When we do not wish to distinguish between the two cases we shall refer to the space as H.

These spaces are Hilbert spaces with the inner product

(16) $$(f, g) = \sum \hat{f}(n)\overline{\hat{g}(n)}[\beta(n)]^2.$$

Let $\hat{f}_k(n) = \delta_{nk}$; in the notation of equation (15) we have $f_k(z) = z^k$. Then $\{f_k\}$ is an orthogonal basis. From (14) it is clear that

(17) $$\|f_k\| = \beta(k).$$

Now consider the linear transformation M_z of multiplication by z on H:

(18) $$(M_z f)(z) = \sum \hat{f}(n)z^{n+1};$$

in other words,

(19) $$(M_z f)\hat{\ }(n) = \hat{f}(n-1) \qquad (\text{all } n)$$

for Laurent series, while

(20) $$\begin{aligned}(M_z f)\hat{\ }(n) &= \hat{f}(n-1) \qquad (n \geqslant 1),\\ &= 0 \qquad\qquad\quad (n = 0),\end{aligned}$$

for power series. Clearly M_z shifts the orthogonal basis $\{f_k\}$.

The linear transformation M_z, defined on finite linear combinations of the basis vectors, need not be bounded. For example, in the power series case, if $[\beta(n)]^2 = n!$ then the series (15) represent entire functions (the so-called Fischer space), and M_z is not bounded: $\|M_z f_n\|^2 = (n+1)\|f_n\|^2$ (see (17)). The study of M_z on this space is related to quantum theory (see [4], [5], [68]).

The following result summarizes what has been said; the proof follows from Proposition 6, equations (13) and (17), and the preceding discussion.

PROPOSITION 7. *The linear transformation* M_z (*on* $L^2(\beta)$ *or* $H^2(\beta)$) *is unitarily equivalent to an injective weighted shift linear transformation (with the weight sequence* $\{w_n\}$ *given below).*

Conversely, every injective weighted shift is unitarily equivalent to M_z (*acting on* $L^2(\beta)$ *or* $H^2(\beta)$, *for a suitable choice of* β).

The relation between $\{w_n\}$ *and* β *is given by the equations*

$$w_n = \beta(n+1)/\beta(n) \qquad (\text{all } n),$$

$$\beta(n) = w_0 \cdots w_{n-1} \qquad (n > 0),$$

$$\beta(0) = 1,$$

$$\beta(-n) = (w_{-1} \cdots w_{-n})^{-1} \qquad (n > 0).$$

COROLLARY. M_z *is bounded if and only if* $\beta(k+1)/\beta(k)$ *is bounded, and in this case*

$$\|M_z^n\| = \sup_k [\beta(k+n)/\beta(k)], \qquad n = 0, 1, 2, \cdots. \tag{21}$$

PROOF. Apply Proposition 2.

For the remainder of this article we shall assume that M_z is a bounded operator.

We have seen that the two points of view summarized by the phrases "weighted shift operator on ordinary l^2" and "ordinary shift operator on weighted l^2" are equivalent. We shall use these two viewpoints interchangeably when convenient, but the greater emphasis will be on the second point of view, since it leads naturally to analytic function theory. There is an abuse of language here; we use the same symbol M_z to represent many different operators (multiplication by z on $L^2(\beta)$ or $H^2(\beta)$, for all β). This should not cause any serious difficulty.

To illustrate the new point of view we reformulate Theorems 1 and 2 (see §2) in the new language.

THEOREM 1′. (a) *The operator* M_z *on* $L^2(\beta)$ *is unitarily equivalent to the operator* M_z *on* $L^2(\beta_1)$ *if and only if there is an integer* k *such that* $\beta_1(n) = \beta(n+k)/\beta(k)$ *for all* n. *Equivalently,* $L^2(\beta) = z^k L^2(\beta_1)$, *and*

$$\|f\|_1 = \|z^k f\|/\|z^k\| \qquad (f \in L^2(\beta_1)),$$

where $\|f\|_1$ *denotes the norm of* f *in* $L^2(\beta_1)$.

(b) *The operators* M_z *on* $H^2(\beta)$ *and* $H^2(\beta_1)$ *are unitarily equivalent if and only if* $\beta = \beta_1$ (*that is,* $H^2(\beta)$ *and* $H^2(\beta_1)$ *have the same elements with the same norm*).

THEOREM 2′. (a) *The following statements are equivalent*:

(i) *The operators* M_z *on* $L^2(\beta)$ *and* $L^2(\beta_1)$ *are similar.*

(ii) *There is an integer* k *and constants* $c, c_1 > 0$ *such that* $c\beta_1(n) \leqslant \beta(n+k)/\beta(k) \leqslant c_1\beta_1(n)$ *for all* n.

(iii) $L^2(\beta) = z^k L^2(\beta_1)$ *and the norms are equivalent* (*that is,* $c\|f\|_1 \leqslant \|z^k f\| \leqslant c_1\|f\|_1$ *for all* f *in* $L^2(\beta_1)$).

(iv) $L^2(\beta) = z^k L^2(\beta_1)$.

(b) *The following statements are equivalent*:

(i) *The operators* M_z *on* $H^2(\beta)$ *and* $H^2(\beta_1)$ *are similar.*

(ii) *There are constants* c, c_1 *such that* $c\beta \leqslant \beta_1 \leqslant c_1\beta$.

(iii) $H^2(\beta) = H^2(\beta_1)$ *and the norms are equivalent.*

(iv) $H^2(\beta) = H^2(\beta_1)$.

Note. The equivalence of (iii) and (iv) is a consequence of the following lemma. A family of linear functionals on a vector space is said be be *separating* if to each nonzero vector there corresponds at least one functional in the family that does not vanish at this vector.

LEMMA. *Let* V *be a real or complex vector space and let* F *be a separating family of linear functionals. Suppose that* V *is complete with respect to two different norms and that the functionals in* F *are continuous with respect to both norms. Then the norms are equivalent.*

PROOF. We define a new norm by adding the two given norms:

$$\|f\| = \|f\|_1 + \|f\|_2 .$$

We claim that V is complete in this norm. Indeed, if $\{f_n\}$ is a Cauchy sequence, then it is also a Cauchy sequence with respect to each of the original norms and so there are elements f, g in V, such that $f_n \longrightarrow f$ in the first norm and $f_n \longrightarrow g$ in the second norm.

Since the functionals in F are continuous in both norms we have $\lambda(f) = \lambda(g)$ for all $\lambda \in F$. Hence $f = g$, and so $f_n \longrightarrow f$ in the new norm.

Thus V is complete with respect to the new norm. Since this norm is greater than each of the original norms it follows that it is equivalent to each of them (by the open mapping theorem) which completes the proof.

4. The commutant. The *commutant* of a set $\mathcal{A}$ of operators is the set of all those operators that commute with each operator in $\mathcal{A}$.

Notation. The commutant of $\mathcal{A}$ is denoted $\mathcal{A}'$; if $\mathcal{A}$ consists of a single operator T it is denoted by $\{T\}'$.

Consider the multiplication of formal Laurent series, $fg = h$:

$$\left(\sum \hat{f}(n)z^n\right)\left(\sum \hat{g}(n)z^n\right) = \sum \hat{h}(n)z^n \tag{22}$$

where

$$\hat{h}(n) = \sum_k \hat{f}(k)\hat{g}(n-k) \qquad (\text{all } n). \tag{23}$$

In the case of formal power series, i.e., $\hat{f}(n) = \hat{g}(n) = 0$ for $n < 0$, the series in (23) become finite sums and so convergence is no problem: The formal power series form an algebra. In the case of general Laurent series we will consider that the product (22) is defined only if all the series (23) are convergent. Recall the notations $H^2(\beta)$ and $L^2(\beta)$ (introduced in §3 after equation (15)).

Notation. $H^\infty(\beta)$ denotes the set of formal power series $\phi(z) = \Sigma\, \hat{\phi}(n)z^n$ $(n \geqslant 0)$ such that $\phi H^2(\beta) \subset H^2(\beta)$.

$L^\infty(\beta)$ denotes the set of formal Laurent series $\phi(z) = \Sigma\, \phi(n)z^n$ $(-\infty < n < \infty)$ such that $\phi L^2(\beta) \in L^2(\beta)$.

Thus $\phi \in L^\infty(\beta)$ means that the product ϕf is defined, and lies in $L^2(\beta)$, for all $f \in L^2(\beta)$. Recall that $f_k(z) = z^k$.

Since f_0 is in $L^2(\beta)$ and $H^2(\beta)$, and $\phi f_0 = \phi$ for all Laurent series ϕ, we see that

$$L^\infty(\beta) \subset L^2(\beta), \qquad H^\infty(\beta) \subset H^2(\beta). \tag{24}$$

Note that for any formal Laurent series ϕ we have

$$z^m\phi(z) = \sum \hat{\phi}(k)z^{k+m} = \sum \hat{\phi}(k-m)z^k \qquad (\text{all } m).$$

Thus, for ϕ in $L^2(\beta)$ or $H^2(\beta)$,

$$(f_m\phi, f_n) = \hat{\phi}(n-m)[\beta(n)]^2 \qquad (\text{all } m). \tag{25}$$

We shall need the concept of the matrix of an operator with respect to an orthogonal (not necessarily normal) basis. Let $\{f_n\}$ be an orthogonal basis for H. A vector $f \in H$ is represented by the sequence $\{\hat{f}(k)\}$, where $f = \Sigma\, \hat{f}(k)f_k$. An operator A is represented by the matrix whose (i, j)th entry is $(Af_j, f_i)/\|f_i\|^2$. Also, if A and B are operators with corresponding matrices (a_{ij}) and (b_{ij}) then AB is represented by the matrix whose (i, j)th term is

$\Sigma_k a_{ik}b_{kj}$, and this series is convergent for all i, j. We omit the details.

Notation. If $\phi \in L^\infty(\beta)$ (or $H^\infty(\beta)$), then the linear transformation of multiplication by ϕ on $L^2(\beta)$ (or on $H^2(\beta)$) will be denoted in both cases by M_ϕ.

PROPOSITION 8. *M_ϕ is a bounded linear transformation.*

PROOF. From (25) we see that the matrix (a_{nk}) of M_ϕ, with respect to the orthogonal basis $\{f_n\}$, is given by $a_{nk} = \hat{\phi}(n-k)$. (We note that when ϕ is a formal power series, then this matrix is lower triangular ($a_{nk} = 0$ for $k > n$).) It is well known that an everywhere defined matrix transformation must be bounded (see, for example, **[18]**).

PROPOSITION 9. *If $\phi, \psi \in L^\infty(\beta)$ (or $H^\infty(\beta)$), then the product $\phi\psi$ is defined and is an element of $L^\infty(\beta)$ (or $H^\infty(\beta)$): $M_\phi M_\psi = M_{\phi\psi}$.*

PROOF. By the previous proposition M_ϕ and M_ψ are bounded operators, whose matrices (with respect to the orthogonal basis $\{f_n\}$) have (i, j)th entries $\hat{\phi}(i-j)$ and $\hat{\psi}(i-j)$, respectively. Hence the product operator $M_\phi M_\psi$ is represented by the product of these two matrices. In particular, the series that arise in the product matrix are all convergent. The (i, j)th entry in the product matrix is

$$\sum_n \hat{\phi}(i-n)\hat{\psi}(n-j) = \sum_k \hat{\phi}(k)\hat{\psi}(i-j-k),$$

and this is precisely the formula for $(\phi\psi)^\wedge(i-j)$, given by (23).

Note. In the formal power series case one knows in advance that products are always defined, and obey the associative law. Hence the result follows without using Proposition 8, by noting that for all $f \in H^2(\beta)$ we have $(\phi\psi)f = \phi(\psi f) \in H^2(\beta)$.

Thus $L^\infty(\beta)$ and $H^\infty(\beta)$ are commutative algebras of operators containing M_z. We now show that the commutant of M_z on $L^2(\beta)$ is $L^\infty(\beta)$, and the commutant of M_z on $H^2(\beta)$ is $H^\infty(\beta)$. (This result is implicit in Proposition 5 of §2.)

THEOREM 3. (a) *If A is an operator on $L^2(\beta)$ that commutes with M_z, then $A = M_\phi$, for some $\phi \in L^\infty(\beta)$.*

(b) *If A is an operator on $H^2(\beta)$ that commutes with M_z, then $A = M_\phi$ for some $\phi \in H^\infty(\beta)$.*

PROOF. (a) and (b). Let $\phi = Af_0$. Then

$$Af_k = AM_z^k f_0 = M_z^k Af_0 = z^k\phi = \phi f_k \qquad (k \geq 0).$$

In case (a) we have further

$$z^{-k}(Af_k) = M_z^{(-k)} Af_k = Af_0 = \phi \qquad (k < 0).$$

Thus if we define λ_n by $\lambda_n(g) = \hat{g}(n)$, then $\lambda_n(Af_k) = \hat{\phi}(n-k)$ for all k. Let $g = \Sigma\, \hat{g}(k) f_k$. Since the series converges in norm we see that $Ag = \Sigma\, \hat{g}(k)(Af_k)$ converges in norm. Hence

$$\lambda_n(Ag) = \sum \hat{g}(k)\lambda_n(Af_k) = \sum \hat{g}(k)\hat{\phi}(n-k),$$

which converges for all n. From (23) this tells us that $A = M_\phi$ as required.

COROLLARY 1. *If T is an injective weighted shift (unilateral or bilateral), then its commutant is a maximal abelian subalgebra of $L(H)$.*

PROOF. This is true for any operator with an abelian commutant.

COROLLARY 2. *If T is an injective unilateral weighted shift, then there does not exist any pair of proper invariant subspaces M_1 and M_2 (not necessarily orthogonal complements) such that $H = M_1 \oplus M_2$.*

PROOF. The existence of such a pair is equivalent to the assertion that T commutes with some idempotent operator other than 0 and I. This would mean that that there would be a formal power series $\phi \in H^\infty(\beta)$ such that $\phi^2 = \phi$. But there are no idempotent formal power series except 0 and f_0.

Question 1. Can an injective unilateral weighted shift have two proper invariant subspaces whose intersection is trivial?

The Beurling theory for the unweighted shift ($w_n \equiv 1$) shows that in this case the intersection of two proper invariant subspaces is nontrivial [**46**, Chapter 7]. Question 8 in §7 is an interesting special case of Question 1. *Added in proof.* These questions have been answered; see the end of this article.

COROLLARY 3. *If T is an injective unilateral weighted shift, then T has no nth roots in $L(H)$ for $n > 1$.*

PROOF. If $A^n = T$, then A commutes with T. But there is no formal power series ϕ such that $\phi^n = z$, if $n > 1$.

We shall use the following notation for the norm of the operator M_ϕ:

(26) $$\|\phi\|_\infty = \|M_\phi\|.$$

We note that $\|\phi\|_2 \leqslant \|\phi\|_\infty$. Indeed, $\|\phi\|_2 = \|M_\phi f_0\|_2 \leqslant \|M_\phi\|\|f_0\|_2 = \|M_\phi\|$, where the subscript 2 denotes the norm in $L^2(\beta)$ (or $H^2(\beta)$).

COROLLARY 4. *Both* $L^\infty(\beta)$ *and* $H^\infty(\beta)$ *are commutative Banach algebras in the norm* (26).

PROOF. Completeness in the norm follows from the fact that the commutant of an operator is always norm closed (in fact, closed in the weak operator topology).

The spaces $H^\infty(\beta)$ and $L^\infty(\beta)$ are (isometric to) conjugate Banach spaces. This follows from the corollary to the following result. First some notation. Let X be a Banach space. For x in X and γ in X^* we denote the value of the functional γ at the vector x by (x, γ). If $\Gamma \subset X^*$, then Γ induces a weak topology on X, which we call the $w(X, \Gamma)$ topology. A net x_α converges to 0 in this topology if $(x_\alpha, \gamma) \longrightarrow 0$ for each γ in Γ.

LEMMA. *Let* X *be a Banach space, and let* $\Gamma \subset X^*$ *be a vector subspace. If the unit ball of* X *is compact in the* $w(X, \Gamma)$ *topology, then there is a Banach space* Y *such that* X *is isometrically isomorphic to* Y^*.

More precisely, Y *can be taken to be the closure of* Γ *in* X^*. *The map of* X *to* Y^* *is the obvious map induced by the pairing of* X *and* X^*.

For the proof see **[89]**.

COROLLARY. *If* X *is a reflexive Banach space and if* T *is a bounded linear operator on* X, *then the commutant of* T, *in the operator norm, is* (*isometrically isomorphic to*) *a conjugate Banach space.*

PROOF. The commutant of T is a weak operator closed subset of the space of all bounded operators on X. Hence its unit ball is compact in the weak operator topology (see [**23**, Exercise VI.9.2]). The result now follows from the lemma if we let Γ denote the set of all weak operator continuous linear functionals on the commutant.

Question 2. Can it happen that there are two nonisomorphic Banach spaces X and Y such that $H^\infty(\beta)$ (or $L^\infty(\beta)$) is isometrically isomorphic both to X^* and to Y^*?

A few words are in order. When we say that a Banach space B "is" a conjugate Banach space there are really two different things that could be meant. First, there exists a Banach space A such that B is isometrically isomorphic to

A^* (the isometric case). Second, B is topologically isomorphic to A^*; that is, there is a bounded invertible linear map of B onto A^* (the isomorphic case). In the first case B can be identified, as a normed space, with A^*. In the second case B can be identified with a space obtained from A^* by re-norming it with an equivalent norm.

To illustrate this we first mention that if B is isometric to A^*, then the unit ball of B must have extreme points (Kreĭn-Milman theorem, plus weak* compactness). Surprisingly, however, if B is merely isomorphic to A^*, then the unit ball of B need not have any extreme points at all. V. Klee **[52]** has given an example of a new norm on l^∞, equivalent to the usual norm, in which the unit ball has no extreme points. If $x = \{x_n\} \in l^\infty$ then one such new norm is given by $\|x\| = \max(\frac{1}{2}\|x\|_\infty, \limsup |x_n|)$. Thus the nonexistence of extreme points in the unit ball of B rules out the possibility that B is isometric to a conjuate Banach space, but leaves open the possibility that B is isomorphic to a conjuate Banach space. However, Bessaga and Pełczyński **[11]** have shown that if X is a Banach space and if X^* is separable, then each bounded, norm-closed convex subset of X^* is the norm closure of the convex hull of its extreme points. Hence a separable Banach space whose unit ball contains no extreme points is not isomorphic to any conjugate Banach space.

Also, a Banach space can be isometric to A^* and to A_1^* where A and A_1 are not isometric (or even isomorphic) to one another. Indeed, if X is a countable compact metric space then $C(X)^*$ is isometric to l^1. On the other hand, $C(X)$ and $C(X_1)$ are isometric if and only is X and X_1 are homeomorphic (see **[23**, V.8.8**]**). Miliutin **[66]** has shown that if X and X_1 are compact metric spaces with uncountably many points, then $C(X)$ and $C(X_1)$ are always isomorphic to one another. Bessaga and Pełczyński **[10]** have shown that when X and X_1 are countable then $C(X)$ and $C(X_1)$ need not be isomorphic (they have given an exact description of when they are isomorphic).

In the isomorphic theory Question 2 has been answered. Consider unweighted shifts ($w_n \equiv 1$). Pełczyński **[74a]** has shown that L^∞ and l^∞ are isomorphic. However their preduals, L' and l', are not isomorphic (L' is not isomorphic to any conjugate space since its unit ball has no extreme points). A similar result holds for H^∞. The proof uses Per Beurling's result **[16a]** that H^∞ is isomorphic to $H^\infty \oplus l^\infty$; see the end of this article.

Several years ago C. E. Rickart asked whether H^∞ is a second conjugate space. This is still unknown. The same question can be asked in our context.

Question 2′. If $H^\infty(\beta)$ (or $L^\infty(\beta)$) isomorphic (or isometric) to a second conjugate Banach space X^{**}?

5. The spectrum. Using Theorem 3 we can now describe the spectrum of a weighted shift. We have already remarked (Corollary 2 to Proposition 1) that the spectrum (and the parts of the spectrum) have circular symmetry.

For any operator A let $\Lambda(A)$ denote the spectrum of A, and let $r(A)$ denote the *spectral radius* of A:

$$r(A) = \sup\{|z|: z \in \Lambda(A)\}.$$

Thus the spectrum lies in the disc $|z| \leqslant r(A)$. We have $r(A) = \lim \|A^n\|^{1/n}$ (see [**34**, Problem 74]).

THEOREM 4. *Let T be a unilateral shift (not necessarily injective). Then the spectrum of T is the disc $|z| \leqslant r(T)$.*

PROOF. First assume that none of the weights is zero. Let λ be in the resolvent set. Since T commutes with $(T - \lambda)^{-1}$, Theorem 3 tells us that $(T - \lambda)^{-1}$ is represented, on an appropriate space $H^2(\beta)$, by M_ϕ, for some formal power series $\phi \in H^\infty(\beta)$. Computing from $(z - \lambda)\phi(z) = 1$ we find that $\hat{\phi}(n) = (-1)/\lambda^{n+1}$ $(n \geqslant 0)$. From (25) we have

$$|\hat{\phi}(k)[\beta(m + k)]^2| = |(M_\phi f_m, f_{m+k})| \leqslant \|M_\phi\|\beta(m)\beta(m + k). \tag{27}$$

Hence $\beta(m + k)/\beta(m) \leqslant |\lambda|^{k+1}\|M_\phi\|$. From (21) we have $\|M_z^k\| \leqslant |\lambda|^k\|\lambda M_\phi\|$. Taking kth roots and letting $k \longrightarrow \infty$ gives $r(T) \leqslant |\lambda|$. Strict inequality must hold since, by circular symmetry, the entire circle $|\lambda| = r(T)$ is in the spectrum. This proves the theorem for injective shifts.

Every shift is the limit in norm of injective shifts $T = \lim T_n$, $r(T_n) \geqslant r(T)$ (just replace the zero weights by small positive numbers). Thus the disc $|z| \leqslant r(T)$ is contained in the spectrum of T_n. But spectra are upper semicontinuous in the norm topology: If G is a neighborhood of the spectrum of T, then G contains the spectrum of T_n for large n (see [**34**, Problem 86]). The result follows.

If an operator A is invertible, we may consider the spectral radius of its inverse. Since $\Lambda(A^{-1}) = [\Lambda(A)]^{-1}$ we have

$$[r(A^{-1})]^{-1} = \inf\{|z|: z \in \Lambda(A)\}.$$

Hence $\Lambda(A)$ lies in the annulus $r(A^{-1})^{-1} \leqslant |z| \leqslant r(A)$.

PROPOSITION 10. *A bilateral weighted shift operator T (represented as M_z on $L^2(\beta)$) is invertible if and only if $\beta(j)/\beta(j + 1)$ is bounded. In this case*

$$\|T^{-n}\| = \sup_j \frac{\beta(j)}{\beta(n+j)} = \left[\inf_j \frac{\beta(n+j)}{\beta(j)}\right]^{-1}, \qquad n = 0, 1, 2, \cdots. \tag{28}$$

Note. This formula is really (21) for negative n.

PROOF. Since $M_z f_m = f_{m+1}$ we have $M_z^{-1} f_m = f_{m-1}$, for all n, where M_z^{-1} is not necessarily bounded but is defined on finite linear combinations of basis vectors. Let $f'_m = f_{-m}$. Then $\{f'_m\}$ is an orthogonal basis that is shifted (forward) by M_z^{-1}. Hence we may represent T^{-1} as multiplication by z on $L^2(\beta')$, where $\beta'(m) = \|f'_m\| = \|f_{-m}\| = \beta(-m)$. From the Corollary to Proposition 7 we see that M_z^{-1} is bounded if and only if $\beta'(m+1)/\beta'(m)$ is bounded, which is equivalent to the condition stated above. In this case we have, from (21),

$$\|T^{-n}\| = \sup_k \frac{\beta'(n+k)}{\beta'(k)} = \sup_k \frac{\beta(-n-k)}{\beta(-k)} = \sup_j \frac{\beta(j)}{\beta(n+j)}, \qquad n = 0, 1, 2, \cdots.$$

The last equality in (28) is immediate.

THEOREM 5. (a) *If T is an invertible bilateral weighted shift, then the spectrum of T is the annulus $[r(T^{-1})]^{-1} \leqslant |z| \leqslant r(T)$.*

(b) *If T is a bilateral weighted shift that is not invertible, then the spectrum is the disc $|z| \leqslant r(T)$.*

PROOF. Assume that T is injective. The general case reduces to this just as in Theorem 4. Represent T as M_z on $L^2(\beta)$, and let λ lie in the resolvent set. Then $(M_z - \lambda)^{-1}$ is M_ϕ, for some $\phi \in L^\infty(\beta)$. From $(z-\lambda)\phi(z) = 1$ we obtain $\hat{\phi}(-1) - \lambda\hat{\phi}(0) = 1$, and

$$\lambda^k \hat{\phi}(k) = \hat{\phi}(0), \qquad \hat{\phi}(-k-1) = \lambda^k \hat{\phi}(-1) \qquad (k \geqslant 0); \tag{29}$$

when $k = \lambda = 0$ we interpret λ^k as 1. Using (25), with n replaced by $m + k$, and (17), we have

$$|\hat{\phi}(k)|\beta(m+k) \leqslant \beta(m)\|M_\phi\| \qquad (\text{all } k, m). \tag{30}$$

Before worrying about whether T is or is not invertible we consider two cases.

Case 1. $\hat{\phi}(0) \neq 0$. If we multiply (30) by $|\lambda|^k$ and apply (29) we obtain

$$|\hat{\phi}(0)|\beta(m+k) \leqslant |\lambda|^k \beta(m)\|M_\phi\| \qquad (k \geqslant 0).$$

Now divide by $\beta(m)$, take the supremum on m, then take the kth root and let $k \longrightarrow \infty$; from (21) we obtain $r(T) \leqslant |\lambda|$. Equality is excluded since by circular symmetry the entire circle $|\lambda| = r(T)$ is in the spectrum.

Case 2. $\hat{\phi}(-1) \neq 0$. In (30) let $k = -n$ $(n \geqslant 1)$, and apply the second equation in (29). We have

$$|\lambda|^{n-1}|\hat{\phi}(-1)|\beta(m-n) \leqslant \beta(m)\|M_\phi\|. \tag{31}$$

First take $n = 1$ and divide by $\beta(m)$, recalling that in case $\lambda = 0$ then 0^0 is interpreted as 1. From the first part of Proposition 10 we see that T is invertible.

Next, for $n \geqslant 1$ divide (31) by $\beta(m-n)$, take the infimum on m, and apply (28). Now take the nth root and let $n \longrightarrow \infty$; we obtain $|\lambda| \leqslant [r(T^{-1})]^{-1}$. As in Case 1 equality is excluded.

At least one of these two cases must hold for each given λ, since $\hat{\phi}(-1) - \lambda\hat{\phi}(0) = 1$. On the other hand they cannot both occur since the conclusions exclude one another. If T is invertible then the two cases together yield the desired conclusion. If T is not invertible, then Case 2 is excluded and Case 1 yields the desired conclusion.

Finally, note that in Case 1 we have $\hat{\phi}(-1) = 0$ and so $\hat{\phi}(n) = 0$ for all $n < 0$ (by (29)); that is, ϕ is a formal power series. Similarly, in Case 2 we have $\hat{\phi}(n) = 0$ $(n > 0)$.

We shall now make a more detailed study of the spectrum of a shift; first we require some concepts from general operator theory. If A is any operator, let

$$m(A) = \inf\{\|Af\|: \|f\| = 1\}$$

denote the lower bound of A. Thus $\|Af\| \geqslant m(A)\|f\|$, for all f. If A is invertible, then $m(A) = \|A^{-1}\|^{-1}$. The following proposition proves a preliminary result; a stronger result is given in Proposition 13.

PROPOSITION 11. *If* $|\lambda| < m(A)$, *then* $A - \lambda$ *is bounded below.*

PROOF. $\|(A-\lambda)f\| \geqslant \|Af\| - \|\lambda f\| \geqslant (m(A) - |\lambda|)\|f\|$.

PROPOSITION 12. $\sup_{n\geqslant 1} [m(A^n)]^{1/n} = \lim_{n\to\infty} [m(A^n)]^{1/n}$.

PROOF. We have $m(AB) \geqslant m(A)m(B)$. Hence

$$m(A^{i+j}) \geqslant m(A^i)m(A^j). \tag{32}$$

This is the crucial inequality. The result now follows from Problem 98, I. Abschnitt, of [75]. For completeness we give the proof.

Fix k. For all n we have $n = kq + r$, $0 \leqslant r < k$, where $q = q(n)$, $r = r(n)$ are functions of n. Note that $\lim q(n)/n = 1/k$. From (32) we have $m(A^n) \geqslant m(A^k)^q m(A)^r$. Hence $\liminf m(A^n)^{1/n} \geqslant m(A^k)^{1/k}$. Since this holds for all k we have

$$\liminf m(A^n)^{1/n} \geqslant \sup m(A^n)^{1/n} \geqslant \limsup m(A^n)^{1/n}$$

and the result follows.

Notation. $r_1(A) = \lim [m(A^n)]^{1/n}$.

If A is invertible, then $r_1(A) = [r(A^{-1})]^{-1}$.

Let $\pi(A)$ denote the *approximate point spectrum* of A (see [**34**, Problem 58]). Thus $0 \in \pi(A)$ if and only if $m(A) = 0$.

PROPOSITION 13. *If* $\lambda \in \pi(A)$, *then* $|\lambda| \geqslant r_1(A)$.

PROOF. Let λ be given, $|\lambda| < r_1(A)$. Thus, for some integer n, $|\lambda|^n < m(A^n)$. By Proposition 11 we see that

$$A^n - \lambda^n = (A^{n-1} + A^{n-2}\lambda + \cdots + \lambda^{n-1})(A - \lambda)$$

is bounded below. Hence $A - \lambda$ is bounded below; that is, $\lambda \notin \pi(A)$.

PROPOSITION 14. *Let* T *be a weighted shift. Then*

$$m(T^n) = \inf_k w_k \cdots w_{k+n-1} \qquad (n = 1, 2, \cdots).$$

If T *is injective and is represented as* M_z *(on* $H^2(\beta)$ *or* $L^2(\beta)$*) then*

$$m(T^n) = \inf_k \frac{\beta(n + k)}{\beta(k)} \qquad (n = 1, 2, \cdots).$$

The proof is elementary and will be omitted.

We state the following theorem without proof.

THEOREM 6 (RIDGE [**79**, p. 350]). *Let* T *be an injective unilateral weighted shift. Then* $\pi(T) = \{\lambda: r_1(T) \leqslant |z| \leqslant r(T)\}$.

Note. Ridge also gives the approximate point spectrum in case some of the weights are zero.

In dealing with bilateral shifts we shall use the following notations.

$$r_1^+(T) = \lim_{n\to\infty} \left[\inf_{j\geqslant 0} \beta(n, j)\right]^{1/n}, \qquad r^+(T) = \lim_{n\to\infty} \left[\sup_{j\geqslant 0} \beta(n, j)\right]^{1/n},$$

$$r_1^-(T) = \lim_{n\to\infty}\left[\inf_{j<0}\beta(-n, j)\right]^{1/n}, \qquad r^-(T) = \lim_{n\to\infty}\left[\sup_{j<0}\beta(-n, j)\right]^{1/n},$$

where $\beta(n, j) = \beta(n + j)/\beta(j)$. The existence of the limit is established just as in Proposition 12. Note that $r_1^-(T) \leqslant r^-(T)$, and $r_1^+(T) \leqslant r^+(T)$. We state the next theorem without proof.

THEOREM 7 (RIDGE [79, p. 352]). *If T is an injective bilateral shift and if $r^-(T) < r_1^+(T)$, then*

$$\pi(T) = \{r_1^- \leqslant |z| \leqslant r^-\} \cup \{r_1^+ \leqslant |z| \leqslant r^+\}.$$

Otherwise,

$$\pi(T) = \Lambda(T) = \{\min(r_1^-, r_1^+) \leqslant |z| \leqslant \max(r^-, r^+)\}.$$

COROLLARY. *Let T be an injective bilateral weighted shift. Then*

$$\max(r^-, r^+) = r(T), \qquad \min(r_1^-, r_1^+) = r_1(T).$$

PROOF. $\pi(T)$ contains the boundary of $\Lambda(T)$ [34, Problem 63]. Apply Theorem 5. (It is also possible to prove the Corollary directly.)

Notation. $\pi_0(A)$ denotes the point spectrum of A (the set of eigenvalues).

If T is an injective unilateral shift with weight sequence $\{w_n\}$, then we define $r_2(T)$ by

$$r_2(T) = \operatorname*{lim\text{-}inf}_{n\to\infty}[w_0 \cdots w_{n-1}]^{1/n}.$$

Thus $r_1(T) \leqslant r_2(T) \leqslant r(T)$.

THEOREM 8. *Let T be an injective unilateral shift. Then*

(i) *$\pi_0(T)$ is empty;*

(ii) $\{0\} \cup \{|z| < r_2(T)\} \subset \pi_0(T^*) \subset \{|z| \leqslant r_2(T)\}$.

Furthermore, all eigenvalues of T^* are simple.

Note. By circular symmetry (Corollary 2 to Proposition 1), one of the two containments in (ii) must be equality.

PROOF. (i) Trivial.

(ii) Assume that λ is an eigenvalue for T^* and that $\sum a_n e_n$ is a corresponding eigenvector. A little calculation shows that $\lambda a_n = w_n a_{n+1}$, for all $n \geqslant 0$. Therefore

$$a_n = a_0\lambda^n/w_0 w_1 \cdots w_{n-1} \qquad (n \geqslant 1). \tag{33}$$

From this we see that the eigenvalues are simple. Further, the calculation works in reverse, and so λ will be an eigenvalue if and only if the sequence $\{a_n\}$ defined by (33) (with $a_0 = 1$) is square summable. We obtain a power series in $|\lambda|$ and the result follows from the Cauchy-Hadamard formula for the radius of convergence. Q.E.D.

If T is an injective bilateral shift with weight sequence $\{w_n\}$ then we define

$$r_2^+(T) = \liminf_{n\to\infty} [w_0 \cdots w_{n-1}]^{1/n}, \qquad r_3^+(T) = \limsup_{n\to\infty} [w_0 \cdots w_{n-1}]^{1/n},$$

$$r_2^-(T) = \liminf_{n\to\infty} [w_{-1} \cdots w_{-n}], \qquad r_3^-(T) = \limsup_{n\to\infty} [w_{-1} \cdots w_{-n}]^{1/n}.$$

Then $r_1^- \leqslant r_2^- \leqslant r_3^- \leqslant r^-$ and $r_1^+ \leqslant r_2^+ \leqslant r_3^+ \leqslant r^+$.

THEOREM 9. *Let* T *be an injective bilateral weighted shift. Then*

(i) *all eigenvalues of* T *and* T^* *are simple*;

(ii) $\{r_3^+ < |z| < r_2^-\} \subset \pi_0(T) \subset \{r_3^+ \leqslant |z| \leqslant r_2^-\}$;

(iii) $\{r_3^- < |z| < r_2^+\} \subset \pi_0(T^*) \subset \{r_3^- \leqslant |z| \leqslant r_2^+\}$;

(iv) *at least one of* $\pi_0(T)$, $\pi_0(T^*)$ *is empty.*

REMARKS. (1) If $r_2^- < r_3^+$ then $\pi_0(T)$ is empty; if $r_2^+ < r_3^-$ then $\pi_0(T^*)$ is empty.

(2) By circular symmetry, one of the containments in (ii), and one of the containments in (iii) must be equality.

PROOF. If λ is an eigenvalue of T and if $f = \Sigma a_n e_n$ is a corresponding eigenvector then a little calculation shows that $a_{n-1}w_{n-1} = \lambda a_n$, for all n. From this we obtain

$$(34) \quad a_n = a_0 w_0 \cdots w_{n-1}/\lambda^n, \qquad a_{-n} = a_0 \lambda^n / w_{-1} \cdots w_{-n} \qquad (n \geqslant 1).$$

From this we see that the eigenvalues are simple. Further, the calculations are reversible and so λ will be an eigenvalue for T if and only if the sequence $\{a_n\}$ defined by (34) (with $a_0 = 1$) is square summable. This leads to two power series, one in λ and one in $1/\lambda$, and the result follows from the formula for the radius of convergence.

The case of T^* is similar.

Finally, assume that λ and μ are eigenvalues of T and T^*, respectively. We wish to show that this is impossible. By what has gone before we must have

$r_3^+ \leqslant |\lambda| \leqslant r_2^-$ and $r_3^- \leqslant |\mu| \leqslant r_2^+$. Hence we have equality throughout and $|\lambda| = |\mu|$. Also, an examination of the series which must converge shows that

$$\sum_1^\infty |w_0 \cdots w_{n-1}|^2/|\lambda|^{2n} < \infty, \qquad \sum_1^\infty |\mu|^{2n}/|w_0 \cdots w_{n-1}|^2 < \infty,$$

which is impossible since $|\lambda| = |\mu|$.

The various numbers $r_i(T)$ are a bit of a nuisance; the following trivial proposition shows that in certain important special cases this nuisance goes away. *Note.* For unilateral shifts we define r_3 just as r_3^+ was defined for bilateral shifts.

PROPOSITION 15. *Let T be a weighted shift with weight sequence $\{w_n\}$. If T is a unilateral shift, and if $w_n \longrightarrow d$, then*

$$r_1 = r_2 = r_3 = r = d.$$

If T is bilateral, if $w_n \longrightarrow d^+$ $(n \longrightarrow \infty)$ and $w_n \longrightarrow d^-$ $(n \longrightarrow -\infty)$, then

$$r_1^- = \cdots = r^- = d^-, \qquad r_1^+ = \cdots = r^+ = d^+.$$

We conclude this section with a word about the numerical range.

If A is any operator let $W(A)$ denote the numerical range and let $w(A) = \sup\{|z|: z \in W(A)\}$ denote the numerical radius. The numerical range is a convex set (see [**34**, Problem 166]).

We shall say that an operator T has circular symmetry if T and $e^{i\theta}T$ are unitarily equivalent for all θ. By Corollary 2 to Proposition 1 all weighted shifts have this property.

For any operator T we define its real and imaginary parts A and B by $A = (T + T^*)/2$, $B = (T - T^*)/2i$. Then A and B are selfadjoint and $T = A + iB$.

The closure of a set K will be denoted by K^-.

PROPOSITION 16. *Let T have circular symmetry. Then*

(i) $\{\lambda: |\lambda| < w(T)\} \subset W(T) \subset \{\lambda: |\lambda| \leqslant w(T)\}$,

(ii) $W(A)^- = W(B)^- = \{t \in \mathbf{R}: -w(T) \leqslant t \leqslant w(T)\}$,

(iii) $w(T) = \|A\| = \|B\|$.

PROOF. (i) The numerical range of T is convex and has circular symmetry.

(ii) $W(A) = \{\mathrm{Re}\,\lambda: \lambda \in W(T)\}$ and similarly for $W(B)$, with $\mathrm{Im}\,\lambda$ in place of $\mathrm{Re}\,\lambda$.

(iii) A and B are selfadjoint and so $w(A) = \|A\|$ and the same for B.

There does not seem to be any simple formula for the numerical radius of a general weighted shift. For the calculation in one particular case of interest, see [**8**, pp. 1053–1054].

6. Analytic structure. In this section we shall discuss the extent to which our formal power series and Laurent series actually represent analytic functions. We shall do this in detail in the case of power series and then indicate the results more briefly for Laurent series. We begin with the notion of a bounded point evaluation.

DEFINITION. If w is a complex number then λ_w denotes the functional of "evaluation at w," defined on polynomials by $\lambda_w(p) = p(w)$.

DEFINITION. w is said to be a *bounded point evaluation* on $H^2(\beta)$ if the functional λ_w extends to a bounded linear functional on $H^2(\beta)$. In this case $f(w)$ will be used to denote $\lambda_w(f)$, for $f \in H^2(\beta)$.

Since the polynomials are dense in $H^2(\beta)$, this is equivalent to the existence of a constant $c > 0$ such that $|\lambda_w(p)| \leqslant c\|p\|_2$ for all polynomials p (the norm denotes the norm in $H^2(\beta)$). Hence there is a unique element $k_w \in H^2(\beta)$ such that for all $f \in H^2(\beta)$ we have

$$f(w) = \lambda_w(f) = (f, k_w) = \sum \hat{f}(n)\overline{\hat{k}_w(n)}[\beta(n)]^2. \tag{35}$$

We call k_w the *reproducing kernel* for $H^2(\beta)$ (associated with the point w). By taking for f the monomial $f_n = z^n$ we obtain

$$\hat{k}_w(n) = \bar{w}^n/[\beta(n)]^2. \tag{36}$$

Thus w is a bounded point evaluation if and only if

$$\|k_w\|^2 = \sum |w|^{2n}/[\beta(n)]^2 < \infty. \tag{37}$$

Note that if p and q are polynomials, then $\lambda_w(pq) = \lambda_w(p)\lambda_w(q)$. Holding p fixed we have, for all $f \in H^2(\beta)$,

$$\lambda_w(pf) = p(w)\lambda_w(f). \tag{38}$$

THEOREM 10. *Let T be an injective unilateral weighted shift operator, represented as M_z on $H^2(\beta)$. Then*:

(i) *w is a bounded point evaluation on $H^2(\beta)$ if and only if $w \in \Pi_0(T^*)$.*

(ii) *If w is a bounded point evaluation and if $f \in H^2(\beta)$, then the power series f converges absolutely at w to the value $\lambda_w(f)$:*

$$\lambda_w(f) = \sum \hat{f}(n)w^n, \qquad \sum |\hat{f}(n)|\,|w|^n < \infty.$$

Furthermore, the disc $\Delta_2(T) = \{|w| < r_2(T)\}$ *is the largest open disc in which all the power series in* $H^2(\beta)$ *converge.*

(iii) *If* $|w| < r(T)$ *and if* $\phi \in H^\infty(\beta)$, *then the power series* ϕ *converges at* w *and*

(39) $$|\phi(w)| \leqslant \|M_\phi\|.$$

Furthermore, this is the largest open disc in which all the power series in $H^\infty(\beta)$ *converge.*

(iv) *If* $\phi \in H^\infty(\beta)$ *and* $f \in H^2(\beta)$ *and* w *is a bounded point evaluation on* $H^2(\beta)$, *then* $\lambda_w(\phi f) = \lambda_w(\phi)\lambda_w(f)$.

(v) *If* $w \in \Pi_0(T^*)$, *then* k_w *is a common eigenvector for all operators commuting with* T^*:

$$M_\phi^* k_w = \overline{\phi(w)} k_w \qquad (\phi \in H^\infty(\beta)).$$

(vi) *If* $|w| = \|T\|$ *then* w *is not a bounded point evaluation.*

(vii) *If the power series* ϕ *represents a bounded analytic function in the disc* $|z| < \|T\|$, *then* $\phi \in H^\infty(\beta)$ *and*

(40) $$\|M_\phi\| \leqslant \sup \{|\phi(z)|: \ |z| < \|T\|\}.$$

PROOF. (i) Let w be a bounded point evaluation on $H^2(\beta)$. From (35) and (38) we have, for all $f \in H^2(\beta)$,

$$(f, M_z^* k_w) = (M_z f, k_w) = w(f, k_w) = (f, \bar{w} k_w).$$

Hence $M_z^* k_w = \bar{w} k_w$ and so $\bar{w} \in \Pi_0(T^*)$. By circular symmetry, $w \in \Pi_0(T^*)$.

Conversely, let $w \in \Pi_0(T^*)$. By circular symmetry, $\bar{w} \in \Pi_0(T^*)$. Let $k \neq 0$ be a corresponding eigenvector: $M_z^* k = \bar{w} k$. Let $\lambda(f) = (f, ck)$, where $c \neq 0$ will be determined later. Then

$$\lambda(zf) = (M_z f, ck) = (f, \bar{w} ck) = w\lambda(f).$$

Hence $\lambda(f_{n+1}) = w\lambda(f_n)$, where $f_n = z^n$. Hence

(41) $$\lambda(f_n) = w^n \lambda(f_0) \qquad (\text{all } n).$$

If $\lambda(f_0) = 0$, then $\lambda = 0$, which is impossible. Hence we may choose c so that $\lambda(1) = 1$. From (41) we then have $\lambda(p) = p(w)$ for all polynomials, and so w is a bounded point evaluation.

(ii) Let $s_n(f) = \Sigma_0^n \hat{f}(k) z^k$ $(n = 0, 1, \cdots)$. These polynomials are the partial sums of the power series f, and converge to f in the norm of $H^2(\beta)$. Hence $\lambda_w(s_n(f)) \to \lambda_w(f)$. But $\lambda_w(p) = p(w)$ for polynomials and so

$$\lim_{n\to\infty} \sum_0^n \hat{f}(k)w^k = \lambda_w(f).$$

Thus the power series converges at w to the right value. To show absolute convergence note first that, by circular symmetry, we have convergence at $|w|$ for all $f \in H^2(\beta)$, and second, if $f \in H^2(\beta)$ then the power series $\Sigma\ |f(n)|z^n$ is also in $H^2(\beta)$. Hence this latter series must converge at $|w|$.

The second part of (ii) follows by noting that the power series f defined by $\hat{f}(n) = 1/(n+1)\beta(n)$ is in $H^2(\beta)$ and has $r_2(T)$ as its exact radius of convergence.

(iii) From (25) we have $|\hat{\phi}(n-m)|\beta(n) \leqslant \|M_\phi\|\beta(m)$ for all n, m. If we divide by $\beta(n)$, and let $n = m + k$, and take the infimum on k, we obtain

$$|\hat{\phi}(k)| \leqslant \|M_\phi\|\,\|T^k\|^{-1}.$$

Hence ϕ converges in the disc $|w| < r(T)$.

Since $H^\infty(\beta)$ is a commutative Banach algebra with identity and evaluation at w is a multiplicative linear functional, this functional must have norm one. This establishes (39).

Finally, if we let $\hat{\phi}(n) = 1/n^2\|T^n\|$ then the resulting power series ϕ is in $H^\infty(\beta)$ (since the series $\Sigma\ T^n/n^2\|T^n\|$ converges in the operator norm) and has radius of convergence $r(T)$.

(iv) From (ii) and (iii) the power series f, ϕ, and ϕf converge absolutely at w to the values $\lambda_w(f)$, $\lambda_w(\phi)$ and $\lambda_w(\phi f)$, respectively. We complete the proof by recalling that the product of two absolutely convergent power (or Laurent) series converges to the product value.

(v) From (iv) we have, for all $f \in H^2(\beta)$,

$$(f, M_\phi^* k_w) = (M_\phi f, k_w) = \phi(w)(f, k_w) = (f, \overline{\phi(w)}k_w)$$

and the result follows.

(vi) We must show that $\Sigma\ \|T\|^{2n}/\beta(n)^2 = \infty$ (see (37)). But this is tirival since $\beta(n) \leqslant \|T^n\| \leqslant \|T\|^n$.

(vii) Our basic tool is the following inequality of von Neumann (see **[34**, Problem 180**]**), valid for all operators on Hilbert space:

$$(42) \qquad \|p(T)\| \leqslant \sup\{|p(z)|: \ |z| < \|T\|\}$$

(here p is any polynomial). This is the desired inequality (40) for the special case when ϕ is a polynomial. If now ϕ is merely a bounded analytic function in $|z| < \|T\|$, then the averages of the partial sums of the power series for

ϕ are bounded by the same bound as ϕ, and converge to ϕ uniformly on compact sets (see, e.g. [**46**, Chapter 2]). Denoting these averages by σ_n, we have

$$\lim_{n\to\infty} \hat{\sigma}_n(k) = \hat{\phi}(k) \qquad (\text{all } k).$$

Thus $(\sigma_n f_j, f_i) \longrightarrow (\phi f_j, f_i)$ as $n \longrightarrow \infty$, for all i, j. From (42) we have

$$\|M_{\sigma_n}\| \leq \sup |\sigma_n(z)| \leq \sup |\phi(z)| \qquad (|z| < \|T\|).$$

This implies weak operator convergence of these operators to M_ϕ. Since $H^\infty(\beta)$ is closed in the weak operator topology, this implies that $\phi \in H^\infty(\beta)$, as desired. (With a little more work it can be shown that σ_n converges to ϕ in the strong operator topology; see Theorem 12 in §8.) Q.E.D.

Notation. If G is an open set in the complex plane then $H^\infty(G)$ denotes the algebra of bounded analytic functions in G with the supremum norm.

If G is a disc or annulus centered at the origin we shall identify the elements of $H^\infty(G)$ with their corresponding power series or Laurent series.

COROLLARY. *If $r(T) = \|T\|$, then the normed algebras $H^\infty(\beta)$ and H^∞ $(|z| < \|T\|)$ are identical.*

PROOF. Apply (iii) and (vii) of the theorem.

Notation. If w is a bounded point evaluation on $H^2(\beta)$, then $H^2_w(\beta)$ denotes the kernel of λ_w (i.e., the functions vanishing at w).

PROPOSITION 17. *$H^2_w(\beta)$ is a closed subspace of codimension one in $H^2(\beta)$, and the polynomials in $H^2_w(\beta)$ are dense in it.*

PROOF. The first part is trivial. For the second, let $f \in H^2_w(\beta)$ and let $\{p_n\}$ be a sequence of polynomials with $p_n \longrightarrow f$. Thus $\lambda_w(p_n) \longrightarrow 0$. Let $q_n = p_n - \lambda_w(p_n)$. Then $\{q_n\}$ is a sequence of polynomials in $H^2_w(\beta)$ converging to f.

PROPOSITION 18. *If w is a bounded point evaluation, then $(z - w)H^2(\beta)$ is a dense subset of $H^2_w(\beta)$. They are equal if and only if $|w| < r_1(T)$.*

PROOF. Clearly $(z - w)H^2(\beta)$ contains all polynomials vanishing at w, and, by Theorem 10(iv), it is a subset of $H^2_w(\beta)$. Density follows from Proposition 17. Next, by Theorem 6, if $|w| < r_1(T)$ then the operator $M_z - wI$ is bounded below and so has a closed range. Also, if $|w| \geq r_1(T)$ (and $|w| \leq r(T)$) then $M_z - wI$ is not bounded below. Hence, since it is one-to-one (Theorem 8), it cannot have a closed range.

We omit the proof of the following corollary.

COROLLARY. *Let* w *be a bounded point evaluation. If* $f \in H^2(\beta)$ *and if* $|w| < r_1(T)$, *then the power series*

$$f_w = (f - f(w))/(z - w) \tag{43}$$

is in $H^2(\beta)$. *If* $|w| \geq r_1(T)$ *then there exists* $f \in H^2(\beta)$ *such that* f_w *is not in* $H^2(\beta)$.

The following result is a reformulation of what has been said; we omit the proof.

THEOREM 11. *Let* T *be an injective unilateral shift. Then for all* w *the operator* $T - wI$ *is one-to-one, and its range*

(i) *is closed and has codimension one if* $|w| < r_1(T)$,

(ii) *is not closed, and its closure has codimension one if* $|w| \geq r_1(T)$ *and* $w \in \Pi_0(T^*)$,

(iii) *is dense but not closed if* $|w| \leq r(T)$ *and* $w \notin \Pi_0(T^*)$,

(iv) *is the whole space if* $|w| > r(T)$.

Further, when T *is represented as* M_z *on* $H^2(\beta)$, *then, for all* w, *the closure of the range of* $M_z - wI$ *coincides with the closure of the set of polynomials vanishing at* w.

PROPOSITION 19. *If* $\phi \in H^\infty(\beta)$ *and if* $|w| < r_1(T)$, *then* $\phi_w \in H^\infty(\beta)$.

Note. ϕ_w is defined as in (43).

PROOF. The operator of multiplication by $\phi - \phi(w)$ has its range contained in $H^2_w(\beta)$. By Theorem 11(i) the operator of multiplication by $(z - w)$ has a bounded inverse (namely, multiplication by $1/(z - w)$ from $H^2_w(\beta)$ to $H^2(\beta)$). Since $\phi_w = [1/(z - w)]\,[\phi - \phi(w)]$, it follows that multiplication by ϕ_w is a bounded operator.

Notation. $L(H^2(\beta))$ denotes the Banach algebra of all bounded operators on $H^2(\beta)$.

PROPOSITION 20. *Let* $\phi \in H^\infty(\beta)$. *Then the spectrum of* ϕ *(as an element of the Banach algebra* $H^\infty(\beta)$*) is the same as the spectrum of* M_ϕ *(as an operator on* $H^2(\beta)$*).*

PROOF. Let b be a complex number. We must compare the invertibility of $\phi - b$ in $H^\infty(\beta)$ with the invertibility of $M_\phi - bI = M_{\phi - b}$ in $L(H^2(\beta))$. Clearly, invertibility in the first sense implies invertibility in the second, since

$H^\infty(\beta) \subset L(H^2(\beta))$. Conversely, let S be a bounded operator on $H^2(\beta)$ that inverts $M_{\phi-b}$. Since M_z commutes with $M_{\phi-b}$ it also commutes with S, and thus $S \in H^\infty(\beta)$. Therefore, the two notions of invertibility coincide.

We now say a few words about the maximal ideal space $\mathcal{M}$ of $H^\infty(\beta)$. Recall that f_1 denotes the power series z.

PROPOSITION 21. *Let* λ *be a multiplicative linear functional of* $H^\infty(\beta)$. *Then the function*

$$\lambda \mapsto \lambda(f_1) \tag{44}$$

is a continuous map of $\mathcal{M}$ *onto* $\Lambda(T)$.

PROOF. By the preceding proposition, $\Lambda(T)$ coincides with the spectrum of f_1 as an element of $H^\infty(\beta)$. This latter spectrum is just the range of the function (44). The continuity is trivial.

DEFINITION. If $w \in \Lambda(T)$, then the inverse image of w in $\mathcal{M}$, under the map (44), is called the *fiber* over w and is denoted $\mathcal{M}_w$. The fiber over w is said to be trivial if it contains only one point.

PROPOSITION 22. *If* $|w| < r_1(T)$, *then* $\mathcal{M}_w$ *is trivial.*

PROOF. Let $\lambda \in \mathcal{M}_w$ and let $\phi \in H^\infty(\beta)$. From Proposition 19 we know that $\phi_w \in H^\infty(\beta)$. Hence

$$\lambda(\phi) - \phi(w) = \lambda(\phi - \phi(w)) = \lambda(f_1 - w)\lambda(\phi_w) = 0.$$

Thus $\lambda(\phi) = \phi(w)$ for all $\lambda \in \mathcal{M}_w$, i.e., the fiber over w is trivial.

Question 3. For which w ($|w| < r(T)$) do we have $\phi_w \in H^\infty(\beta)$ for all $\phi \in H^\infty(\beta)$? For which w is the fiber over w trivial?

If the fiber over w is trivial for all $|w| < r(T)$, then this disc is homeomorphically embedded in the maximal ideal space of $H^\infty(\beta)$.

Question 4 (*Corona conjecture*). If $r(T) > 0$ and if the fiber over w is trivial for all $|w| < r(T)$, then is this disc dense in the maximal ideal space of $H^\infty(\beta)$?

This is known to be true when $\beta = 1$ (see **[16]**, **[17]**, or Chapter 12 of **[24]**). It is also true in the strictly cyclic case (see Corollary 1 to Proposition 31 in §9). Just as in the special case $\beta = 1$ the question is equivalent to the following: If $f_1, \cdots, f_n$ are in $H^\infty(\beta)$ and if

$$|f_1(z)| + \cdots + |f_n(z)| \geq \delta > 0 \qquad (|z| < r(T)),$$

then do there exist $g_1, \cdots, g_n \in H^\infty(\beta)$ such that $\Sigma f_i g_i = 1$? (See [**24**, Chapter 12], or the last corollary in Chapter 10 of [**46**].)

From the discussion preceding Question 2 in §4 we know that $H^\infty(\beta)$ is (isometric to) a conjugate Banach space. This suggests the following problem.

Question 5. What is the relation between the weak* topology and the weak operator topology on $H^\infty(\beta)$?

See the discussion following Question 8 in §7 for an interesting special case of this problem. The two topologies coincide in the strictly cyclic case (§9).

Question 6. Discuss the spectrum of M_ϕ $(\phi \in H^\infty(\beta))$. What is the approximate point spectrum? If $r(T) = 0$, can the spectrum of M_ϕ contain more than one point?

It is easy to see that if $r(T) > 0$, if $\phi \in H^\infty(\beta)$, and if we regard ϕ as a bounded analytic function in the disc $|z| < r(T)$ (see Theorem 10 (iii)), then the closure of the range of ϕ lies in the spectrum of M_ϕ. It does not seem to be known if they must always coincide. They do coincide in two cases: when $r(T) = \|T\|$ (see the Corollary to Theorem 10), and in the strictly cyclic case (see Corollary 1 to Proposition 31 in §9).

An interesting special case occurs when $\beta(n) = (n+1)^{1/2}$. Then $H^2(\beta)$ consists of the analytic functions in the unit disc with a finite Dirichlet integral (that is, the derivative of the function is square integrable with respect to Lebesgue measure on the open disc). The problem now takes the form: If $\phi H^2(\beta) \subset H^2(\beta)$ and if $|\phi(z)| \geqslant d > 0$ for $|z| < 1$, do we have $\phi H^2(\beta) = H^2(\beta)$? *Added in proof.* L. Carleson has proved this; see the end of the article.

If $r(T) = 0$, and if M_z is strictly cyclic, then $\Lambda(M_\phi) = \hat{\phi}(0)$, for $\phi \in H^\infty$ (see Corollary 1 to Proposition 31 in §9).

We now turn briefly to the case of injective bilateral shifts. By a *Laurent polynomial* we mean a terminating Laurent series, that is, finite linear combination of the vectors $\{f_n\}$ $(-\infty < n < \infty)$, where $f_n(z) = z^n$. These polynomials are dense in $L^2(\beta)$. Bounded point evaluations (bpe) are defined as before, with "polynomial" replaced by "Laurent polynomial." ($w = 0$ is always excluded since Laurent polynomials cannot be evaluated at the origin.) We have the following analogue of Theorem 10.

THEOREM 10′. *Let T be an injective bilateral shift operator, represented as M_z on $L^2(\beta)$. Then:*

(i) *w is a bpe on $L^2(\beta)$ if and only if $w \in \Pi_0(T^*)$.*

(ii) *If w is a bpe and if $f \in L^2(\beta)$, then the Laurent series f con-*

verges absolutely at w *to the value* $\lambda_w(f)$. *Furthermore, the annulus* $\{r_3^- < |w| < r_2^+\}$ *is the largest open annulus in which all the Laurent series in* $L^2(\beta)$ *converge.*

(iii) (a) *If* T *is invertible, if* $r(T^{-1})^{-1} < |w| < r(T)$, *and if* $\phi \in L^\infty(\beta)$, *then the Laurent series* ϕ *converges at* w *and*

$$|\phi(w)| \leqslant \|M_\phi\|. \tag{39'}$$

Furthermore, this is the largest open annulus in which all the Laurent series in $H^\infty(\beta)$ *converge.*

(b) *If* T *is not invertible and if* $\phi \in L^\infty(\beta)$ *then* $\hat{\phi}(n) = 0$ *for* $n < 0$ (*that is,* ϕ *is a formal power series*). *If* $|w| < r(T)$, *then the power series* ϕ *converges at* w, *and* (39′) *holds. Furthermore, this is the largest open disc in which all the series in* $L^\infty(\beta)$ *converge.*

(iv) *If* $\phi \in L^\infty(\beta)$, $f \in L^2(\beta)$, *and* w *is a bpe, then* $\lambda_w(\phi f) = \lambda_w(\phi)\lambda_w(f)$. *Thus,* λ_w *is a multiplicative linear functional on* $L^\infty(\beta)$.

(v) *If* $w \in \Pi_0(T^*)$, *then* k_w *is a common eigenvector for all operators commuting with* T^*.

$$M_\phi^* k_w = \overline{\phi(w)} k_w \qquad (\phi \in L^\infty(\beta)).$$

(vi) *The point* w *is not a bounded point evaluation if* $|w| = \|T\|$, *or if* $w = 0$, *or if* $|w| = 1/\|T^{-1}\|$ (*in case* T *is invertible*).

(vii) (a) *If* T *is not invertible and if* ϕ *is a bounded analytic function in the disc* $|z| < \|T\|$, *then* $\phi \in L^\infty(\beta)$ *and*

$$\|M_\phi\| \leqslant \sup\{|\phi(z)|: \ |z| < \|T\|\}.$$

(b) *If* T *is invertible, if* $\|T^{-1}\|^{-1} < \|T\|$, *and if* ϕ *is bounded and analytic in the annulus* $\|T^{-1}\|^{-1} < |z| < \|T\|$, *then* $\phi \in L^\infty(\beta)$ *and*

$$\|M_\phi\| \leqslant c \sup\{|\phi(z)|: \ \|T^{-1}\| < |z| < \|T\|\},$$

where the constant c *depends only on the two numbers* $\|T\|$, $\|T^{-1}\|$, *and not on* ϕ.

(c) *If* T *is invertible and if* $\|T^{-1}\|^{-1} = \|T\|$, *then* $\phi \in L^\infty(\beta)$ *if and only if* ϕ *is a bounded measurable function on the circle* $|z| = \|T\|$, *with Lebesgue measure. In this case*

$$\|M_\phi\| = \operatorname{ess\,sup}\{|\phi(z)|: \ |z| = \|T\|\}.$$

PROOF. Parts (i), (ii), (iv), and (v) are proved just as in Theorem 10.

(iii) (a) If $\phi \in L^\infty(\beta)$ let $\phi = \phi_1 + \phi_2$ where

$$\phi_1(z) = \sum_0^\infty \hat{\phi}(n)z^n, \qquad \phi_2(z) = \sum_{-\infty}^{-1} \hat{\phi}(n)z^n.$$

Then just as in the proof of (iii) of Theorem 10 one shows that ϕ_1 converges for $|z| < r(T)$, and ϕ_2 converges for $|z| > r(T^{-1})^{-1}$. Also, (39′) is established as before.

(b) From Proposition 10 we know that $\inf \{\beta(j+1)/\beta(j): -\infty < j < \infty\} = 0$. From (25) (with $n = m - 1$) we have

$$|\hat{\phi}(-1)| \leqslant \|M_\phi\| \beta(m)/\beta(m-1) \qquad (-\infty < m < \infty).$$

Hence $\hat{\phi}(-1) = 0$, for all $\phi \in L^\infty(\beta)$. Multiplying by z we have $\phi(-2) = 0$, etc. Thus we may identify ϕ with ϕ_1 and the result follows as in (a) above.

In the proof of (vi) recall that from (28) we have $\beta(-n) \leqslant \|T^{-n}\| \leqslant \|T^{-1}\|^n$ for $n \geqslant 0$.

(vii) (a) Same proof as for Theorem 10(vii).

(b) Here we have a new difficulty. The proof in the unilateral case (and in (a) above) was based on von Neumann's inequality (42). Now we would require the following analogue: If T is invertible and if $\|T^{-1}\|^{-1} < \|T\|$, then

$$\|p(T)\| \leqslant \sup \{|p(z)|: \|T^{-1}\|^{-1} < |z| < \|T\|\},$$

for all Laurent polynomials p. Unfortunately, this is false. (This follows from Corollary 1 to Theorem 9 of **[97]**.) However, we shall show that the inequality can be rescued if we put a suitable constant on the right side. We require the following result. (In the proof we use the fact that a bounded analytic function in an annulus has radial boundary values on both boundary components. In the application that we shall make of this result we will only be considering Laurent polynomials.)

LEMMA. *Let* $0 < r < R$ *be given, and let* G *denote the annulus* $\{r < |z| < R\}$. *Let* ϕ *be a bounded analytic function in* G, $|\phi| \leqslant M$. *Then* $\phi = \phi_1 + \phi_2$, *where* ϕ_1 *is a bounded analytic function in* $|z| < R$, ϕ_2 *is a bounded analytic function in* $r < |z|$, *and* $\phi_2(\infty) = 0$. *Furthermore,*

$$|\phi_1(z)| \leqslant [1 + r/(R^2 - r^2)^{1/2}]M, \qquad |z| < R,$$

$$|\phi_2(z)| \leqslant [1 + R/(R^2 - r^2)^{1/2}]M, \qquad |z| > r.$$

PROOF. Let

$$\phi(z) = \sum_{-\infty}^{\infty} a_n z^n = \sum_{0}^{\infty} + \sum_{-\infty}^{-1} = \phi_1(z) + \phi_2(z), \qquad z \in G.$$

Then ϕ_1 is analytic in $|z| < R$, ϕ_2 is analytic in $|z| > r$, and $\phi_2(\infty) = 0$. Also, ϕ_2 is bounded in a neighborhood of the circle $|z| = R$, and hence ϕ_1 is bounded in $|z| < R$. Similarly, ϕ_2 is bounded in $|z| > r$. Furthermore

$$|\phi_1(re^{i\theta})| \leqslant \sum_{0}^{\infty} |a_n| r^n \leqslant \left(\sum_{-\infty}^{\infty} |a_n|^2 R^{2n} \right)^{1/2} R/(R^2 - r^2)^{1/2}$$

$$= \left(\frac{1}{2\pi} \int_{-\pi}^{\pi} |\phi(Re^{i\theta})|^2 \, d\theta \right)^{1/2} R/(R^2 - r^2)^{1/2} \leqslant MR/(R^2 - r^2)^{1/2}.$$

Hence

$$|\phi_2(re^{i\theta})| = |\phi(re^{i\theta}) - \phi_1(re^{i\theta})| \leqslant M[1 + R/(R^2 - r^2)^{1/2}].$$

The proof of the other inequality is similar.

PROPOSITION 23. *Let T be an invertible operator on Hilbert space, and let $R = \|T\|$, $r = \|T^{-1}\|^{-1}$, and assume that $r < R$. Then*

$$\|p(T)\| < \left[2 + \left(\frac{R + r}{R - r} \right)^{1/2} \right] \sup \{ |p(z)| : \ r < |z| < R \}, \tag{45}$$

for all Laurent polynomials p.

PROOF. We have $p = p_1 + p_2$, as in the lemma. We may assume that $|p| \leqslant 1$ in the annulus. Then from the lemma

$$\|p(T)\| \leqslant \|p_1(T)\| + \|p_2(T)\| \leqslant \sup_{|z|<R} |p_1(z)| + \sup_{|z|>r} |p_2(z)|$$

$$\leqslant 2 + (R + r)/(R^2 - r^2)^{1/2}$$

and the result follows.

Note. We have used von Neumann's inequality for $p_1(T)$; for $p_2(T)$ we have used the inequality obtained from von Neumann's inequality by replacing z by $1/z$ and T by T^{-1}.

The remainder of the proof of Theorem 10′(vii)(b) follows the lines of the proof of Theorem 10(vii).

(vii) (c) Let $c = \|T\| = \|T^{-1}\|^{-1}$. If $\{w_n\}$ is the weight sequence for T, it follows that $|w_n| = c$, for all n. Hence T is unitarily equivalent to the shift with the constant weight sequence c. Thus $\beta(n) = c^n$ (all n), and $L^2(\beta)$ coincides with $L^2(d\theta/2\pi c)$ on $|z| = c$. The result follows.

Question 7. Find the smallest constant on the right side of (45). In particular, if $R = 1$ is this constant bounded (as a function of r)? Also, in Theorem $10'$(vii)(b) can we take $c = 1$?

A corollary, analogous to the corollary to Theorem 10, is valid. The analogues of Propositions 17 through 21 inclusive, hold for injective bilateral shifts; the condition $|w| < r_1(T)$ in Propositions 18 and 19 is replaced by the condition: w is not in the approximate point spectrum of T. Finally, the analogues of Questions 3–6 are open.

We close this section with the following result.

PROPOSITION 24. *Let $\{\phi_n\}$ be a sequence in $H^\infty(\beta)$ (or in $L^\infty(\beta)$). Then ϕ_n converges to 0 in the weak operator topology if and only if $\{\|\phi_n\|\}$ is bounded, and $\hat{\phi}_n(k) \longrightarrow 0$ $(n \longrightarrow \infty)$ for all k.*

PROOF. If $\phi_n \longrightarrow 0$ (WOT), then the norms are bounded (by the principle of uniform boundedness) and $\hat{\phi}_n(k) \longrightarrow 0$ (by (25)). Conversely, if these two conditions hold, then $(\phi_n f_i, f_n) \longrightarrow 0$ $(n \longrightarrow \infty)$ for all i, j. Hence $(\phi_n f, g) \longrightarrow 0$ for all f, g in a dense set (finite combinations of basis vectors). The weak convergence follows from this since the $\{\phi_n\}$ are bounded.

7. **Hyponormal and subnormal shifts.** A bounded linear operator A on Hilbert space H is said to be *hyponormal* if $A^*A - AA^* \geqslant 0$. It is said to be *subnormal* if it is the restriction of a normal operator to an invariant subspace. More precisely, this means that there is a Hilbert space K, containing H as a closed subspace, and a normal operator B on K having H as an invariant subspace such that $A = B|H$. The proof of the following result is elementary and will be omitted.

LEMMA. *Let T be a weighted shift operator (unilateral or bilateral) with (nonnegative) weight sequence $\{w_n\}$. Then T is hyponormal if and only if*

$$w_n \leqslant w_{n+1} \qquad (\textit{all } n). \tag{46}$$

A bilateral shift is normal if and only if $\{w_n\}$ is constant (that is, $w_n = c$ for all n); a unilateral shift is never normal.

Notes. (1) In case T is injective and is represented as M_z on $H^2(\beta)$ or $L^2(\beta)$ then (46) becomes

$$[\beta(n)]^2 \leqslant \beta(n-1)\beta(n+1) \qquad (\text{for all } n). \tag{46}$$

(2) The condition $\beta(n+1) \leqslant \beta(n)$ (all n) is equivalent to $\|T\| \leqslant 1$,

which is equivalent to $w_n \leqslant 1$ (all n).

We now discuss subnormal shifts. First, the unilateral case. Let μ be a probability measure on the interval $[0, R]$ $(R > 0)$, such that the point R is in the support of μ (that is, $\mu([R - \epsilon, R]) > 0$ for all $\epsilon > 0$). We consider the probability measure $d\sigma = d\mu(r)d\theta/2\pi$ on the disc $|z| \leqslant R$ (see [**35**, p. 896]). The operator M_z of "multiplication by z" on the Hilbert space $K = L^2(d\sigma)$ is a bounded normal operator. Let H denote the closed subspace spanned by $\{z^n\}$ $(n \geqslant 0)$. Then H is an invariant subspace for M_z, and M_z restricted to H is a subnormal operator. It is clear that the vectors $\{z^n\}$ form an orthogonal basis for H, and thus $M_z|H$ is a unilateral shift.

Let

$$[\beta(n)]^2 = \int r^{2n}\, d\mu(r) \qquad (n \geqslant 0). \tag{47}$$

A little calculation shows that if p is a polynomial then

$$\|p\|^2 = \frac{1}{2\pi} \iint |p(re^{it})|^2\, d\theta\, d\mu(r) = \sum |\hat{p}(n)|^2 [\beta(n)]^2.$$

Since H is the completion of the polynomials in this norm, it may be identified with $H^2(\beta)$, where $\beta(n)$ is defined in (47). Thus the sequence $\{\beta(n)\}$ is the sequence of moments of even index of a probability measure on $[0, R]$. We may consider the transformation $\psi(r) = r^2$, from $[0, R]$ to $[0, R^2]$. This induces a measure μ_1 on $[0, R^2]$ defined by $\mu_1(E) = \mu(\psi^{-1}(E))$. The formula (47) now becomes

$$[\beta(n)]^2 = \int r^n\, d\mu_1(r) \qquad (n \geqslant 0); \tag{48}$$

thus the sequence $\{\beta(n)\}$ is the sequence of moments of a probability measure on $[0, R^2]$. Note that (46)$'$ is satisfied, since in our present situation M_z is subnormal, and hence is hyponormal. Or we may derive it directly from the Cauchy inequality:

$$\left(\int r^n\, d\mu_1(r)\right)^2 = \int r^{(n-1)/2} r^{(n+1)/2}\, d\mu_1(r)$$
$$\leqslant \left(\int r^{n-1}\, d\mu_1(r)\right)\left(\int r^{n+1}\, d\mu_1(r)\right).$$

PROPOSITION 25. *Let T be an injective unilateral shift, represented as M on $H^2(\beta)$. Then T is subnormal if and only if $\{\beta(n)\}$ is the moment sequence of some probability measure μ_1 on $[0, \|T\|^2]$, with $\|T\|^2$ in the support of μ_1.*

PROOF. See [**35**, pp. 895–896].

The measure μ related to μ_1 as in (47), (48) will be called the measure associated with T.

Let $\Delta(T)$ denote the open disc $|z| < r(T)$.

If f is analytic in a neighborhood of $|z| = r$ and if $0 < p < \infty$, let

$$M_p(f, r) = \left(\frac{1}{2\pi}\int_0^{2\pi} |f(re^{i\theta})|^p d\theta\right)^{1/p}$$

and let

$$M_\infty(f, r) = \max_\theta |f(re^{i\theta})|.$$

PROPOSITION 26. *Let the injective unilateral shift T be hyponormal. Then*

(i) $r_1(T) = r_2(T) = r(T) = \|T\|$.

(ii) $H^\infty(\beta)$ *coincides, as a normed algebra, with the bounded analytic functions in* $\Delta(T)$ *with the supremum norm.*

Now assume that T is subnormal, with associated measure μ.

(iii) *If* $\mu(\{\|T\|\}) = 0$, *then* $H^2(\beta)$ *consists of all those power series f that converge in* $\Delta(T)$ *and for which*

$$\|f\|^2 = \int [M_2(f, r)]^2 d\mu(r) < \infty. \tag{49}$$

(iv) *If μ has mass at the point* $\|T\|$ *then* $H^2(\beta)$ *consists of all those power series f that converge in* $\Delta(T)$ *and for which*

$$\sup\{M_2(f, r)\colon\ 0 \leqslant r < \|T\|\} < \infty. \tag{49$'$}$$

PROOF. (i) The first two equalities follow from Proposition 15; the last is trivial since we have $\sup w_n = \lim w_n$.

(ii) This is a special case of the Corollary to Theorem 10.

(iii) A calculation shows that $[M_2(f, r)]^2 = \Sigma\,|f(n)|^2 r^{2n}$. Hence (49) is equivalent to $\Sigma\,|\hat{f}(n)|^2\beta(n)^2 < \infty$.

(iv) We have $c_1\|T\|^n \leqslant \beta(n) \leqslant c\|T\|^n$ for all n and for suitable constants c, c_1. Hence $H^2(\beta)$ consists of those power series f for which $\Sigma\,|\hat{f}(n)|^2\|T\|^{2n} < \infty$; this is equivalent to (49)$'$.

Let T be a subnormal unilateral shift of norm one. By the preceding proposition, $H^2(\beta)$ can be identified with those analytic functions in the unit disc that are square integrable with respect to a probability measure. We shall say that the sequence of points $\{z_n\}$, $|z_n| < 1$, is a *zero set* for $H^2(\beta)$ if there is an $f \in H^2(\beta)$ that vanishes at all the points $\{z_n\}$ but does not vanish identically.

In this case the set of all $f \in H^2(\beta)$ that vanish on the sequence $\{z_n\}$ is a proper closed invariant subspace of $H^2(\beta)$. Question 1 of §4 thus suggests the following problem.

Question 8. Is the union of two zero sets for $H^2(\beta)$ again a zero set for $H^2(\beta)$?

This is unknown even for the Bergman space: $d\mu(r) = 2rdr$, $\beta(n)^2 = (n+1)^{-1}$. *Added in proof*. This has been settled negatively; see the end of this article.

Question 5 in the previous section asks for the relation between the weak operator topology on $H^\infty(\beta)$ and the weak* topology. For subnormal shifts this question takes a more concrete form.

Indeed, let T be an injective unilateral subnormal shift, with $\|T\| = 1$. Then $H^\infty(\beta)$ is the space H^∞ of all bounded analytic functions in the unit disc. We may view H^∞ as being a subspace of $L^\infty(d\sigma)$, where $d\sigma = d\mu(r)d\theta/2\pi$. It can be shown that H^∞ is a weak* closed subspace. Hence H^∞ is (isometric to) the conjugate space of a quotient space of $L'(d\sigma)$. A net $\{\phi_\alpha\}$ in H^∞ converges weak* to 0 if and only if $\lim \int \phi_\alpha f\, d\sigma = 0$ for all $f \in L'$.

For the sake of definiteness let us consider the special case when $H^2(\beta)$ is the Bergman space (those analytic functions in the unit disc that are square integrable with respect to Lebesgue area). Then a net (ϕ_α) in H^∞ converges to 0 in the weak operator topology if and only if

$$(\#) \qquad \lim\,(\phi_\alpha g, h) = \lim \int \phi_\alpha g \bar{h}\, d\sigma = 0$$

for all g, h in $H^2(\beta)$ (now $d\sigma$ denotes Lebesgue measure on the open unit disc). Clearly if $\phi_\alpha \longrightarrow 0$ (w*), then $\phi_\alpha \longrightarrow 0$ (WOT), but it is not known if the converse is valid. Note that in (#) we may actually take h arbitrary in $L^2(d\sigma)$.

We could also view H^∞ as being a (weak* closed) subspace of the boundary space $L^\infty(d\theta)$. This leads to the same weak* topology on H^∞ (see [**82**, 4.1, 4.2, 4.5 on pp. 251–253; 4.23, 4.24 on pp. 261–263]). If we view H^∞ as operating on the Hardy space H^2, then it can be shown that the weak operator topology coincides with the weak* topology. Indeed, if $f \in L^1$ and if $|f| \geqslant d > 0$ a.e., then $|f| = g\bar{g}$ for some g in H^2. Hence $f = g\bar{g}u$, where $|u| = 1$. Thus $f = g\bar{h}$, where $g \in H^2$ and $h \in L^2$. We claim that every f in L^1 is the sum of two functions that are bounded away from 0. Indeed, if f is given, let $v = 1$ when $f = 0$, and let $v = f/|f|$ when $|f| > 0$. Then $f = (f + v) - v$, and $|f + v| = |f| + |v| \geqslant 1$, and $|v| = 1$ a.e. Thus weak operator convergence implies weak* convergence.

There is a well-known theorem (due to C. R. Putnam [**76**]) that if two normal operators are similar, then they are unitarily equivalent (see also [**34**, Problem 152]). Sarason gave the following example to show that this is no longer true for subnormal operators (see [**50**, p. 14]).

Let $d\mu(r) = 2rdr$ and $d\nu(r) = 4r^3dr$ on $0 \leqslant r \leqslant 1$. Then $\beta(n)^2 = 1/(n+1)$ in the first case, and $2/(n+2)$ in the second. The corresponding shift operators are subnormal and are similar (Theorem 2b) but are not unitarily equivalent (Theorem 1b).

For the purposes of the following question we say that the space $H^2(\beta)$ is subnormal, or hyponormal, if the corresponding shift operator is subnormal, or hyponormal.

Question 9. If $H^2(\beta)$ is hyponormal, is it equal, as a set, to the union of all the subnormal spaces contained within it?

In considering this question we may assume that the shift operator on $H^2(\beta)$ has norm 1. It follows that $\beta(n)$ is decreasing. Thus $H^2(\beta)$ contains the usual Hardy space H^2 of the unit disc, so that $H^2(\beta)$ really does contain some subnormal spaces.

In considering subnormal bilateral shifts we have the following result whose proof will be omitted (see [**41**]).

From the discussion preceding Proposition 11 recall that $m(T)$ denotes the lower bound of T on the unit sphere. If T is an injective bilateral shift then it has dense range, and so either $m(T) = 0$ or T is invertible and $m(T) = \|T^{-1}\|^{-1}$.

PROPOSITION 27. *Let T be an injective bilateral shift, represented as M_z on $L^2(\beta)$. Then T is subnormal if and only if there is a probability measure μ on the interval $m(T) \leqslant r \leqslant \|T\|$ such that*

$$\beta(n)^2 = \int r^{2n}\, d\mu(r) \qquad (-\infty < n < \infty).$$

REMARKS. If $m(T) = 0$, then the measure μ must be such that the integrals on the right are finite for all negative n. In particular, μ must have no mass at the origin.

Also, we could replace μ be a measure μ_1 on $[m(T)^2, \|T\|^2]$ which would have $\beta(n)^2$ as its nth moment (instead of $2n$, as above).

Finally, just as in Proposition 26 it can be shown that $r_1^- = \cdots = r^- = m(T)$, and $r_1^+ = \cdots = r^+ = \|T\|$. Also, we can describe the normed algebra $L^\infty(\beta)$ precisely.

If $m(T) = \|T\|$, then $L^\infty(\beta)$ coincides with $L^\infty(ds)$, where s denotes the normalized Lebesgue measure on $|z| = \|T\|$. In this case $L^2(\beta) = L^2(ds)$.

If $m(T) < \|T\|$ then $L^\infty(\beta)$ coincides, as a normal algebra, with the bounded analytic functions in the annulus $m(T) < |z| < \|T\|$. One could describe $L^2(\beta)$ by considering separately the four cases determined by whether μ does or does not put mass at the two points $m(T)$ and $\|T\|$.

8. Algebras generated by shifts. Let A be a bounded operator on Hilbert space H. By the *algebra* generated by A we mean the closure, in the weak operator topology, of the polynomials in A. By the *rational algebra* generated by A we mean the closure, in the weak operator topology, of the rational functions in A (the poles of the rational functions must be disjoint from the spectrum of A). These two algebras are abelian, and are contained in the commutant of A.

One could replace the weak operator topology by other topologies, for example, by the strong operator topology or by the uniform operator topology (that is, the norm topology). The first of these would not lead to anything new: A vector subspace of $L(H)$ is closed in the weak operator topology if and only if it is closed in the strong operator topology (see [**23**, Corollary VI.1.5, p. 477]). The second would lead, in general, to smaller algebras; we shall not discuss them here except for a few passing comments.

We shall show that for injective unilateral and injective noninvertible bilateral shifts the algebra generated by the shift coincides with the commutant, while for injective invertible bilateral shifts the rational algebra (in fact, the algebra generated by T and T^{-1}) coincides with the commutant.

Let $\Gamma = \{|w| = 1\}$.

Let H denote either $H^2(\beta)$ or $L^2(\beta)$.

For $f \in H$ and $w \in \Gamma$ we define f_w by $f_w(z) = f(wz)$, that is

$$\hat{f}_w(n) = w^n \hat{f}(n) \qquad (\text{all } n). \tag{50}$$

Note that this is NOT the same as the definition in equation (43) of §6. Clearly $\|f_w\| = \|f\|$. For fixed f it is easy to see that the map $w \mapsto f_w$ is continuous from Γ into H.

PROPOSITION 28. *Let* $\phi \in H^\infty(\beta)$ *or* $L^\infty(\beta)$ *and* $w \in \Gamma$. *Then*

(i) $\|\phi_w\|_\infty = \|\phi\|_\infty$;

(ii) *for fixed* ϕ *the map* $w \mapsto \phi_w$ *is continuous from* Γ *into* $L(H)$ *in the strong operator topology.*

PROOF. (i) Let $f \in H$. One verifies that $(\phi f)_w = \phi_w f_w$, and that $(f_a)_b = f_{ab}$. Hence

$$\|\phi f\|_2 = \|(\phi f)_w\|_2 = \|\phi_w f_w\|_2 \leqslant \|\phi_w\|_\infty \|f\|_2 .$$

Thus $\|\phi\|_\infty \leqslant \|\phi_w\|_\infty$. Applying this inequality to $\psi = \phi_w$ we have $\|\psi\| \leqslant \|\psi_a\|$. Taking $a = \bar{w}$ the result follows.

(ii) We recall the notion of the matrix of an operator with respect to an orthogonal basis, and the fact that the (n, m)th entry of the matrix of M_ϕ with respect to the basis $\{f_n\}$ is $\hat{\phi}(n - m)$ (see §4, the proof of Proposition 8 and the preceding discussion). Then

$$\|\phi f_m - \phi_w f_m\| = \left\| \sum_n (1 - w^{n-m})\hat{\phi}(n - m) f_n \right\|$$

which tends to zero as $w \longrightarrow 1$, for fixed m. This together with (i) above implies the result.

Let $f \in H$ and $\phi \in H^\infty(\beta)$ (or $L^\infty(\beta)$). The strong operator continuity of ϕ_w allows us to define the vector-valued Riemann integral

$$\int \phi_w p(w) f\, ds \tag{51}$$

where $ds = |dw|/2\pi$ is the normalized Lebesgue measure on Γ and p is continuous on Γ. More generally, (51) will exist as an abstract Lebesgue integral whenever p is integrable (we shall not require this case). In either case we have

$$\left(\int \phi_w p(w) f\, ds,\ g\right) = \int p(w)(\phi_w f,\ g)\, ds \tag{52}$$

and

$$\left\| \int \phi_w p(w) f\, ds \right\|_2 \leqslant \|\phi\|_\infty \|f\|_2 \int |p|\, ds. \tag{53}$$

We define the corresponding operator-valued integral by

$$\left(\int \phi_w p(w)\, ds\right) f = \int \phi_w p(w) f\, ds \tag{54}$$

and from (53) we have

$$\left\| \int \phi_w p(w)\, ds \right\|_\infty \leqslant \|\phi\|_\infty \int |p|\, ds. \tag{55}$$

A *trigonometric polynomial* is a function of the form $p(w) = \Sigma_{-\infty}^{\infty} \hat{p}(k) w^k$ $(w \in \Gamma)$ where only finitely many coefficients are different from zero. Thus it is just the restriction of a Laurent polynomial to Γ.

If $\phi \in H^\infty(\beta)$ or $L^\infty(\beta)$ we shall identify ϕ with the operator M_ϕ of multiplication by ϕ on H.

PROPOSITION 29. *If $\phi \in H^\infty(\beta)$ (or $L^\infty(\beta)$) and if p is a trigonometric polynomial, then $\phi * p \in H^\infty(\beta)$ (or $L^\infty(\beta)$) and*

(56) $$\int \phi_w p(\bar{w})\, ds = M_{\phi * p}$$

where

(57) $$(\phi * p)(z) = \sum \hat{\phi}(h)\hat{p}(h)z^k.$$

PROOF. It is enough to consider the case $p(w) = w^k$. Let ψ denote the left side of (56). Thus ψ is an operator, and we must show that it coincides with the operator (of multiplication by) $\hat{\phi}(k)z^k$. For this it will be sufficient to show that these two operators have the same matrix entries with respect to the orthogonal basis $\{f_n\}$. We have

$$(\hat{\phi}(k)z^k f_m, f_n) = \hat{\phi}(k)(f_{m+k}, f_n).$$

On the other hand from (25) we have

$$(\psi f_m, f_n) = \int \bar{w}^k(\phi_w f_m, f_n)\, ds = \int \bar{w}^k w^{n-m}\hat{\phi}(n-m)\beta(n)^2\, ds$$

$$= \begin{cases} \hat{\phi}(k)\beta(n)^2 & \text{if } n = k, \\ 0 & \text{otherwise.} \end{cases}$$

This completes the proof.

Notation.

$$s_n(\phi) = \sum_{|k| \leqslant n} \hat{\phi}(k)z^k \qquad (n \geqslant 0),$$

$$\sigma_n(\phi) = \frac{S_0(\phi) + \cdots + S_n(f)}{n+1} \qquad (n \geqslant 0).$$

Here ϕ is any power series or Laurent series, $s_n(\phi)$ are the partial sums, and $\sigma_n(\phi)$ the averages (or Cesàro means) of the partial sums of ϕ.

Let $K_n(w)$ $(w \in \Gamma)$ denote the Fejer kernel:

(58) $$K_n(w) = \sum_{-n}^{n} \left(1 - \frac{|k|}{n+1}\right) w^k = \frac{1}{n+1}\left(\frac{\sin(n+1)\theta/2}{\sin\theta/2}\right)^2.$$

(See [**98**, Chapter III, §3].) We have $K_n \geqslant 0$ and $\int K_n\, ds = 1$.

We turn now to the main result of this section.

THEOREM 12. *If* $\phi \in H^\infty(\beta)$ *or* $L^\infty(\beta)$ *then*

(i) $\sigma_n(\phi) \in H^\infty(\beta)$ *or* $L^\infty(\beta)$,

(ii) $\|\sigma_n(\phi)\| \leqslant \|\phi\|$,

(iii) $\sigma_n(\phi) \longrightarrow \phi$ *in the strong operator topology.*

PROOF. From (57) and Proposition 29 we have $\sigma_n(\phi) = \int \phi_w K_n(\bar{w})\, ds$. Assertions (i) and (ii) are immediate from this. A little calculation shows that the column vectors in the matrix for $\sigma_n(\phi)$ with respect to the orthogonal basis $\{f_n\}$ converge in the norm of H to the corresponding column vectors in the matrix for ϕ. In other words, $\sigma_n(\phi)f_k$ converges to ϕf_k in norm, for all k. This together with (ii) implies (iii).

COROLLARY. (a) *Let T be an injective shift. If T is unilateral, or bilateral but not invertible, then each operator that commutes with T is the limit, in the strong operator topology, of a sequence of polynomials in T.*

(b) *If T is bilateral and invertible, then each operator that commutes with T is the limit, in the strong operator topology, of a sequence of Laurent polynomials in T.*

Thus we see that if the injective shift T (unilateral or bilateral) is not invertible, then the algebra generated by the shift coincides with the commutant.

If T is invertible then the commutant coincides not with the algebra generated by T, but rather with the rational algebra (in fact, with the algebra generated by T and T^{-1}).

Question 10. What are the coefficient multipliers on $L^\infty(\beta)$ and on $H^\infty(\beta)$?

We regard L^∞ and H^∞ as sequence spaces (Laurent or Taylor coefficients). A sequence $\{\lambda_n\}$ is said to be a coefficient multiplier on a space A of sequences, if $\{\lambda_n a_n\} \in A$ whenever $\{a_n\} \in A$. It can be shown that if A is a Banach space and if the coordinate functionals are bounded, then the coefficient multipliers are a subset of l^∞ (the bounded sequences).

One can define $\int \phi_w\, d\mu(w)$ for $\phi \in L^\infty(\beta)$ (or $H^\infty(\beta)$) and μ any complex-valued bounded Borel measure on the unit circle Γ. Using this it can be shown that each Fourier-Stieltjes sequence $\{\hat{\mu}(n)\}_{-\infty}^{\infty}$ is a multiplier on $L^\infty(\beta)$, and each sequence $\{\mu(n)\}_0^\infty$ is a multiplier on $H^\infty(\beta)$. Here $\hat{\mu}(n) = \int w^n\, d\mu(w)$ (all n). For the spaces $L^\infty(ds)$ (Lebesgue measure on Γ) and H^∞ (bounded analytic functions in the unit disc) these are the only coefficient multipliers (see [98, Chapter IV, Theorem 11.4]). On the other hand, for some shifts we have $H^\infty = H^2$ (or $L^\infty = L^2$) and so every bounded sequence is a coefficient multiplier (see §9).

By the *analytic projection* we mean the map which sends each two-sided sequence $\{a_n\}_{-\infty}^{\infty}$ into the corresponding one-sided sequence $\{a_n\}_0^\infty$.

Question 11. For which bilateral shifts is the analytic projection a bounded operator on $L^\infty(\beta)$?

In view of Theorem 10′(iii)(b) in §6 it is certainly bounded if the shift is not invertible. On the other hand it is not bounded when $\beta \equiv 1$ (see [**98**, Chapter VII, equations (2.2) and (2.3)]).

9. Strictly cyclic shifts. An algebra A of operators on H is called *strictly cyclic* if there exists a vector f_0 such that $\mathsf{A}f_0 = H$. Such a vector is called a *sttrictly cyclic vector* for A. Recall that f_0 is said to be a *cyclic vector* for an algebra A if $\mathsf{A}f_0$ is dense in H.

PROPOSITION 30. *If* A *is a norm-closed strictly cyclic algebra containing the identity operator, then every cyclic vector for* A *is strictly cyclic.*

PROOF. Let $\rho\colon \mathsf{A} \longrightarrow H$ be the map $\rho(A) = Af_0$ $(A \in \mathsf{A})$, where f_0 is a given strictly cyclic vector. This map is bounded, and by the open mapping theorem the image of each open set is an open set. In particular, if $U \subset \mathsf{A}$ denotes the open ball of radius one about the identity operator, then $\rho(U)$ is open.

Now let f_1 be a cyclic vector. There exists $A \in \mathsf{A}$ such that $Af_1 \in \rho(U)$. Hence $Af_1 = Bf_0$ for some $B \in U$. But the elements of U are invertible and so we have $B^{-1}Af_1 = f_0$. It follows that f_1 is a strictly cyclic vector.

Now let A be an abelian strictly cyclic algebra (ASCA), and let f_0 be a strictly cyclic vector, $\|f_0\| = 1$. We shall review the basic facts about these algebras.

First we note that A is a *maximal* abelian algebra. Indeed, if B is an operator that commutes with all the operators in A, then let A_0 be an operator in A such that $A_0f_0 = Bf_0$ (such an A_0 exists by the definition of strictly cyclic algebra). By commutativity, $A_0(Af_0) = B(Af_0)$ for all $A \in \mathsf{A}$, and so $B = A_0$, i.e., $B \in \mathsf{A}$.

It follows that A is a closed subset of $\mathsf{L}(H)$ in the weak operator topology, and hence also in the norm topology. In particular A is complete in the operator norm, and thus is a commutative Banach algebra with identity.

It follows also that an element $A_0 \in \mathsf{A}$ is invertible in A if and only if it is invertible in $\mathsf{L}(H)$. Indeed, one direction is trivial since $\mathsf{A} \subset \mathsf{L}(H)$. For the converse, assume that there exists $B \in \mathsf{L}(H)$ such that $A_0B = BA_0 = I$. It follows that B commutes with every operator that commutes with A_0. Indeed

$$BA = BAA_0B = BA_0AB = AB.$$

In particular, B commutes with each operator in A, and so $B \in \mathsf{A}$.

Next, let $\rho\colon \mathsf{A} \longrightarrow H$ be the map $\rho(A) = Af_0$ $(A \in \mathsf{A})$.

LEMMA. *The map* ρ *is one-to-one and onto. There exists a constant* c *such that*

$$\|Af_0\| \leqslant \|A\| \leqslant c\|Af_0\| \qquad (A \in \mathsf{A}).$$

PROOF. By assumption ρ is onto. To show that ρ is one-to-one it is enough to show that ρ has trivial kernel. Assume that $\rho(A_0) = 0$. Then $0 = A(A_0 f_0) = A_0(Af_0)$ for all $A \in \mathsf{A}$. Hence $A_0 = 0$.

Finally, $\|Af_0\| \leqslant \|A\|$ and so ρ is continuous. The existence of the constant c follows from the open mapping theorem. Q.E.D.

Thus ρ establishes a Banach space isomorphism between the Banach algebra A and the Hilbert space H. Therefore A admits an equivalent norm in which it is a Hilbert space. However it is not possible for the operator norm on A to be a Hilbert norm (except in the trivial case when A is the complex numbers). This was first shown by L. Ingelstam **[48]**; see M. F. Smiley **[90]** for a simpler proof. They show that if a real Hilbert space is also an algebra with identity, and if the identity has norm 1 and the norm of a product is less than or equal to the product of the norms (i.e., the Hilbert space is a Banach algebra with identity) then it must be isomorphic to the reals, complexes, or quaternions. The case of a complex Hilbert algebra can be reduced to the real case by restricting the scalars to be real, and by taking the real part of the inner product. The conclusion then is that a complex Hilbert space–Banach algebra must be the complex numbers.

Since A and H are isomorphic as Banach spaces, the linear functionals on A may be identified with the elements of H:

$$\lambda_f(A) = (Af_0, f) \qquad (f \in H, \quad A \in \mathsf{A}).$$

Of course this representation of the conjugate space of A depends on the particular strictly cyclic vector that has been chosen.

Using this representation of the conjugate space it can be shown that the weak operator topology on A coincides with the weak topology of A as a Banach space. It follows from the lemma that the strong operator topology on A coincides with the norm topology.

We are now ready to apply the foregoing results to shift operators. We consider first unilateral shifts. If the shift is not injective (this is the same as saying that at least one weight is equal to 0) then it cannot have a cyclic vector and a fortiori it cannot be strictly cyclic. Indeed, let T be a unilateral weighted shift with weight sequence $\{w_n\}$. The range of T is the span of the vectors $\{w_n e_{n+1}\}$ $(n \geqslant 0)$. Thus the codimension of the range equals the number of zero weights, plus one. In particular, if at least one weight is zero then the codimension is at least 2. But then there cannot be a cyclic vector f, since $\{T^n f\}$ $(n \geq 0)$ can span a space of dimension at most one greater than the dimension of the range of T.

Thus in studying strictly cyclic unilateral shifts we may assume that the shift is injective, and hence that it can be represented as M_z on $H^2(\beta)$. We recall from Theorem 12 of §8 that the (weakly closed) algebra generated by M_z is $H^\infty(\beta)$. Also, if f is any power series then $s_n(f)$ denotes its nth partial sum.

PROPOSITION 31. *Let T be an injective unilateral shift represented as M_z on $H^2(\beta)$. Then T is strictly cyclic if and only if $H^2(\beta) = H^\infty(\beta)$ (that is, these two sets have the same elements). In this case we have*

(1) $c\|f\|_\infty \leq \|f\|_2 \leq \|f\|_\infty$ $(f \in H^2(\beta))$,

(2) $r_2(T) = r(T)$,

(3) $s_n(f) \longrightarrow s(f)$ *in operator norm* $(f \in H^\infty(\beta))$,

(4) $\Sigma\, r(T)^{2n}/\beta(n)^2 < \infty$,

(5) $\Sigma\, |\hat{f}(n)| r(T)^n \leq c'\|f\|_2$ $(f \in H^2(\beta))$.

PROOF. Since the power series 1 is always a cyclic vector for M_z, it follows from Proposition 30 that M_z is strictly cyclic if and only if 1 is a strictly cyclic vector, that is, if and only if $H^\infty(\beta) \cdot 1 = H^2(\beta)$. In this case (1) follows from the first lemma in this section.

(2) This follows from Theorem 10(ii), (iii) of §6.

(3) This follows from (1) since we always have norm convergence in $H^2(\beta)$.

(4) It will be sufficient to prove that evaluation at each point w, $|w| = r(T)$, is a bounded point evaluation on $H^2(\beta)$ (see (37) at the beginning of §6). Let w_0 be given, $|w_0| = r(T)$, and let p be a polynomial. From Theorem 10(iii), (39), and from (1) above, we have

$$|p(w_0)| \leq \|M_p\| \leq \|p\|_2/c,$$

and the result follows.

(5) This follows from (4) by the CBS inequality (Cauchy-Buniakowsky-Schwarz).

COROLLARY 1. *If $H^2(\beta) = H^\infty(\beta)$, then the maximal ideal space of $H^\infty(\beta)$ is the closed disc $\Delta(T)^- = \{|z| \leq r(T)\}$, and the spectrum of each element of $H^\infty(\beta)$ is precisely its range on $\Delta(T)^-$.*

PROOF. If λ is a multiplicative linear functional on $H^\infty(\beta)$ then let $w_1 = \lambda(f_1)$ (where f_1 denotes the power series z, as in §3). Hence $\lambda(p) = p(w_1)$ for all polynomials. Since λ is bounded and the polynomials are dense it follows that w_1 is a bounded point evaluation on $H^2(\beta)$ and so $|w_1| \leqslant r(T)$.

Conversely, evaluation at each point of $\Delta(T)^-$ is a multiplicative linear functional on $H^\infty(\beta)$ (see part (4) of the proposition, and equation (37) in §6). The result follows.

COROLLARY 2. *If* $H^2(\beta) = H^\infty(\beta)$ *then a (closed) subspace is invariant under* M_z *if and only if it is an ideal. Further, the functions in any proper invariant subspace all have a common zero in the closed disc* $\Delta(T)^-$.

PROOF. If M is an invariant subspace and if A is any operator in the algebra generated by M_z, then $AM \subset M$. Hence $H^\infty M \subset M$, and so M is an ideal. The converse is trivial.

If M is a proper ideal then it must lie in a maximal ideal. By Corollary 1 this means that the functions in M have a common zero.

COROLLARY 3. *If* $H^\infty(\beta) = H^2(\beta)$ *then* f *is a cyclic vector for* M_z *if and only if* f *never vanishes on* $\Delta(T)^-$.

PROOF. From Proposition 30, f is cyclic if and only if it is strictly cyclic, that is, if and only if it is invertible in $H^\infty(\beta)$. The result now follows from Corollary 1.

Each of the properties (1)–(5) of Proposition 31 is necessary for M_z to be strictly cyclic. Property (1) is sufficient. Indeed, $H^\infty(\beta)$ is always a dense vector subspace of $H^2(\beta)$ (it contains the polynomials). If (1) holds then $H^\infty(\beta)$ is a closed subspace of $H^2(\beta)$ and so they must coincide.

Property (2) is not sufficient for strict cyclicity. Consider, for example, the classical case $\beta(n) \equiv 1$. We shall give an example later in this section to show that (4) is not sufficient. Since (4) implies (5), it follows that (5) is not sufficient. The problem remains open for (3).

Question 12. If $s_n(f) \longrightarrow f$ in operator norm for all $f \in H^\infty(\beta)$, must we have $H^2(\beta) = H^\infty(\beta)$?

Question 13. If T is a strictly cyclic injective unilateral shift, must we have $r_1(T) = r(T)$? (For the definition of $r_1(T)$ see §5, just before Proposition 13.)

PROPOSITION 32. *Let* T *be an injective unilateral shift represented as* M_z *on* $H^2(\beta)$. *Then* T *is strictly cyclic if and only if*

$$\sum_{n=0}^{\infty}\left|\sum_{k=0}^{n} a(k)b(n-k)\frac{\beta(n)}{\beta(k)\beta(n-k)}\right|^2 < \infty \tag{59}$$

for all $a, b \in l^2$.

A necessary condition for this to hold is that

$$\sup_{k,m}\frac{\beta(k+m)}{\beta(k)\beta(m)} < \infty. \tag{60}$$

A sufficient condition for (59) *to hold is that*

$$\sup_{n}\sum_{k=0}^{n}\left(\frac{\beta(n)}{\beta(k)\beta(n-k)}\right)^2 < \infty. \tag{61}$$

If the weight sequence $\{w_n\}$ *for* T *is monotone decreasing, then* (61) *is necessary as well as sufficient for* (59) *to hold.*

PROOF. From Proposition 31, T will be strictly cyclic if and only if $fg \in H^2(\beta)$ for all $f, g \in H^2(\beta)$. This will be the case if and only if

$$\sum_{n=0}^{\infty}\left|\sum_{k=0}^{n}\hat{f}(k)\beta(k)\hat{g}(n-k)\beta(n-k)\frac{1}{\beta(k)\beta(n-k)}\right|^2\beta(n)^2 < \infty.$$

This is the same as (59). Furthermore, by Proposition 31 if (59) holds, then there is a constant c such that the left side of (59) is actually less than or equal to

$$\frac{1}{c^2}\sum |a_k|^2 \sum |b_n|^2. \tag{62}$$

To show that (60) must hold, let n_0 and k_0 be given, with $n_0 \geqslant k_0$. Define a, b by

$$a(k_0) = 1, \qquad a(k) = 0 \qquad (k \neq k_0),$$

$$b(n_0 - k_0) = 1, \qquad b(n) = 0 \qquad (n \neq n_0 - k_0).$$

Then $\|a\| = \|b\| = 1$ in l^2. Let $\beta(n, k) = \beta(n)/\beta(k)\beta(n-k)$. From (62) we have

$$1/c^2 \geqslant \left|\sum_{k=0}^{n_0} a(k)b(n_0-k)\beta(n_0, k)\right|^2 = \beta(n_0, k_0)^2.$$

The boundedness follows.

The sufficiency of (61) follows from (59) by applying the Cauchy inequality.

For the remainder of the proof see Theorem 3.2 of [**51**].

REMARKS. If (60) holds, then a sufficient condition for strict cyclicity is that there exist a nonnegative integer j such that (61) holds for $n \geqslant 2j$, $k \geqslant j$, that is

$$\text{(61)}' \qquad \sup_{n \geqslant j} \sum_{k=j}^{n-j} \left(\frac{\beta(n)}{\beta(k)\beta(n-k)} \right)^2 < \infty.$$

Indeed, using (60) this implies (61).

COROLLARY 1. *If* $\{\lambda_k\}$ *and* $\{\mu_m\} \in l^2$ *and if*

$$\text{(63)} \qquad \beta(k+m)/\beta(k)\beta(m) \leqslant \lambda_k + \mu_m,$$

then M_z *is strictly cyclic.*

PROOF. Condition (61) is satisfied.

COROLLARY 2. *If* M_z *is strictly cyclic as an operator on* $H^2(\beta)$ *and if* β^* *satisfies*

$$\text{(64)} \qquad \beta^*(k+m)/\beta^*(k)\beta^*(m) \leqslant c\beta(k+m)/\beta(k)\beta(m) \qquad (\text{all } k, m),$$

then M_z *is strictly cyclic on* $H^2(\beta^*)$.

PROOF. This is immediate from (59).

COROLLARY 3. *If* T *is a strictly cyclic unilateral weighted shift with weight sequence* $\{w_n\}$, *and if* $a_0 \geqslant a_1 \geqslant \cdots > 0$, *then the shift with weight sequence* $\{a_n w_n\}$ *is also strictly cyclic.*

PROOF. The shift with weights $\{a_n w_n\}$ can be represented as M_z on $H^2(\beta^*)$, where $\beta^*(n) = \beta(n) a_0 a_1 \cdots a_{n-1}$ (see Proposition 7 in §3). The result now follows from the previous corollary.

REMARK. The monotonicity of the sequence $\{a_n\}$ could be replaced by the weaker condition

$$\text{(65)} \qquad \alpha(k+m)/\alpha(k)\alpha(m) \leqslant c \qquad (\text{all } n, m),$$

where $\alpha(0) = 1$, $\alpha(n) = a_0 a_1 \cdots a_{n-1}$ $(n > 0)$. In particular this will hold if the sequence $\{a_n\}$ is eventually monotone decreasing.

LEMMA. *If* $\{a_n\}$ *is positive, decreasing, and if* $\Sigma\, a_n < \infty$, *then* $na_n \longrightarrow 0$.

PROOF. Let $\epsilon > 0$ be given. For large n we have

$$na_{2n} \leqslant a_{n+1} + \cdots + a_{2n} < \epsilon.$$

The result follows.

In particular, if a_n is decreasing and square summable, then $a_n \leqslant c/n^{1/2}$.

COROLLARY 4. *If T is an injective unilateral shift whose weight sequence $\{w_n\}$ is monotone decreasing and in l^p for some $p < \infty$, then T is strictly cyclic.*

PROOF. Choose j such that the sequence $v_n = w_n \cdots w_{n+j-1}$ is square summable. Let $n \geqslant 2j$, and $j \leqslant k \leqslant n - j$. Then

$$\beta(n) = \beta(k)(w_k \cdots w_{n-j-1})v_{n-j}, \qquad \beta(n-k) = (w_j \cdots w_{n-k-1})v_0.$$

Hence

$$\beta(n)/\beta(k)\beta(n-k) \leqslant v_{n-j}/v_0 \leqslant c/(n-j)^{1/2}.$$

Thus (61)$'$ is satisfied, and the result follows.

The following definition will be important when we study invariant subspaces in §10.

Let T be an injective unilateral weighted shift $Te_n = w_n e_{n+1}$ $(n \geqslant 0)$. Let $E_n = \text{span}\ \{e_k\colon\ k \geqslant n\}$.

DEFINITION. T is *strongly strictly cyclic* if $T|E_n$ is strictly cyclic $(n = 0, 1, \cdots)$.

Of course $T|E_n$ is again an injective unilateral weighted shift with weight sequence $\{w_n, w_{n+1}, \cdots\}$. If T is represented as M_z on $H^2(\beta)$, then $T|E_n$ can be represented as M_z on $H^2(\beta_n)$, where $\beta_n(k) = \beta(n+k)/\beta(n)$ $(k \geqslant 0)$.

Recall that an operator A is said to be bounded below if $\|Af\| \geqslant \delta\|f\|$ for all f and for some $\delta > 0$ (see the definition of $m(A)$ in §5, preceding Proposition 11).

PROPOSITION 33. *If T is a strictly cyclic injective unilateral shift, and if T is bounded below, then T is strongly strictly cyclic.*

PROOF. Let T be represented as M_z on $H^2(\beta)$. Then $w_n = \beta(n+1)/\beta(n) \geqslant \delta > 0$. $T|E_1$ is respresented as M_z on $H^2(\beta_1)$. Since

$$\frac{\beta_1(n)}{\beta_1(k)\beta_1(n-k)} = \frac{\beta(n+2)}{\beta(k+1)\beta(n-k+1)} \cdot \frac{\beta(1)\beta(n+1)}{\beta(n+2)}$$

$$\leqslant c\frac{\beta(n+2)}{\beta(k+1)\beta(n-k+1)},$$

from Proposition 32 we see that $T|E_1$ is strictly cyclic. The general case, $T|E_n$, is similar.

REMARK. If $T|E_1$ is strictly cyclic then T itself is strictly cyclic. This is immediate from the relation between β and β_1.

We now give two results which follow from the definition of strong strict cyclicity.

COROLLARY 1. *If T is a strongly strictly cyclic unilateral weighted shift with weight sequence $\{w_n\}$, and if the positive sequence $\{a_n\}$ is eventually decreasing then the shift with weight sequence $\{a_n w_n\}$ is also strongly strictly cyclic.*

PROOF. Choose N such that $a_N \geqslant a_{N+1} \geqslant \cdots$. By Corollary 3 to Proposition 32 each of the shifts $T|E_n$ $(n = N, N + 1, \cdots)$ is strictly cyclic. The cases $n \leqslant N$ are handled by the remark preceding the statement of this corollary.

COROLLARY 2. *If T is an injective unilateral shift with weight sequence that is monotone decreasing and in l^p for some $p < \infty$, then T is strongly strictly cyclic.*

PROOF. Apply Corollary 4 to Proposition 32 to each of the shifts $T|E_n$ $(n = 0, 1, \cdots)$.

We turn to some examples.

EXAMPLE 1. $\beta(n) = (n + 1)^\alpha$, $\alpha > ½$. Then

$$\left[\frac{\beta(n)}{\beta(k)\beta(n-k)}\right]^2 = \left[\frac{n+1}{(k+1)(n-k+1)}\right]^{2\alpha} < \left[\frac{1}{k+1} + \frac{1}{n-k+1}\right]^{2\alpha}$$
$$\leqslant c_\alpha[1/(k+1)^{2\alpha} + 1/(n-k+1)^{2\alpha}].$$

(The second inequality follows from the fact that the function $f(x) = x^a$ $(x > 0)$ is convex when $a \geqslant 1$. Hence $[(x + y)/2]^a \leqslant (x^a + y^a)/2$.) From (61) and from Proposition 33 we see that M_z is strongly strictly cyclic.

In particular with $\alpha = 1$ we see that the shift with weights $w_n = (n + 2)/(n + 1)$ is strongly strictly cyclic.

PROPOSITION 34. *If $w_n(n + 1)/(n + 2)$ is eventually decreasing, then the unilateral shift with weight sequence $\{w_n\}$ is strongly strictly cyclic.*

PROOF. This follows from the previous example (with $\alpha = 1$) and the preceding Corollary 1, if we let $a_n = w_n(n + 1)/(n + 2)$.

EXAMPLE 2. If $w_n = 1/\log(n+2)$ for all n, or if $w_n = 1/\log\log(n+3)$ for all n, then the shift with weight sequence $\{w_n\}$ is strongly strictly cyclic. The same is true for higher order iterated logarithms. This follows from Proposition 34.

Question 14. Is every strictly cyclic shift strongly strictly cyclic?

This is unknown even if we assume that the weight sequence is decreasing.

We now present an example showing that Property 4 of Proposition 31 does not imply strict cyclicity.

EXAMPLE 3. Let $\{w_n\}$ satisfy (i) $w_n \longrightarrow 1$, (ii) $w_n \geqslant (n+2)/(n+1)$ $(n = 0, 1, \cdots)$. From (i) we have $r(T) = 1$, and from (ii) we have $\Sigma\, 1/(\beta(n))^2 < \infty$. Therefore property (4) of Proposition 31 is satisfied, and hence the power series in $H^2(\beta)$ all converge absolutely on the boundary of the spectral disc $|z| \leqslant r(T)$. Nonetheless it is possible to choose the sequence $\{w_n\}$ so that (60) in Proposition 32 is not satisfied, and hence the shift determined by this weight sequence is not strictly cyclic.

Let $u_n = (n+2)/(n+1)$. We choose an increasing sequence of positive integers $\{n_j\}$ such that $2n_j < n_{j+1}$. Let

$$w_k = u_k \qquad (0 \leqslant k < n_1),$$

$$w_k = u_1 \qquad (n_1 \leqslant k < 2n_1),$$

$$w_k = u_k \qquad (2n_1 \leqslant k < n_2),$$

$$w_k = u_2 \qquad (n_2 \leqslant k < 2n_2),$$

etc. As always $\beta(n) = w_0 w_1 \cdots w_{n-1}$. Hence for $k \leqslant n_1$ we have $\beta(k) = k+1$. Choose n_1 to be the first positive integer that satisfies the inequality $u_1^k > \beta(k)$.

For $2n_1 < k \leqslant n_2$ we have $\beta(k) < (k+1)u_1^{n_1}$. Let n_2 be the smallest integer, greater than $2n_1$, such that $u_2^k > 2(k+1)u_1^{n_1}$. Thus $u_2^{n_2} > 2\beta(n_2)$. Continue in this way, in general $u_j^{n_j} > j\beta(n_j)$. Then

$$\beta(2n_j)/\beta(n_j)^2 > j \qquad (j = 1, 2, \cdots)$$

which completes the construction.

Note that a similar construction will work with $w_n \longrightarrow 0$. However, the weight sequence is not monotone.

Question 15. If $w_n \searrow 1$ and if $\Sigma\, 1/\beta(n)^2 < \infty$, must T be strictly cyclic? If $w_n \searrow 0$, must T be strictly cyclic? strongly strictly cyclic?

We now say a few words about bilateral shifts.

PROPOSITION 35. *A bilateral shift is never strictly cyclic.*

PROOF. By definition, the algebra generated by an operator is the closure, in the weak operator topology, of the polynomials in the operator. It follows from Theorem 12 of §8 that the algebra A generated by a bilateral shift coincides with the set of all those formal Laurent series in $L^\infty(\beta)$, whose coefficients of negative index all vanish. In other words, A is the class of all those formal power series that are in $L^\infty(\beta)$.

Assume now that there is a strictly cyclic vector g_0. Then for each n $(-\infty < n < \infty)$ there exists $\phi_n \in \mathsf{A}$ such that $\phi_n g_0 = z^n$. Then $z\phi_n = \phi_{n+1}$ for all n. From $\phi_0 = z\phi_{-1}, \phi_{-1} = z\phi_{-2}$, etc. we see that

$$\phi_0 = z^n \phi_{-n} \qquad (n > 0).$$

But ϕ_{-n} is a formal power series, and so the first n coefficients of ϕ_0 must vanish, for each n, which is a contradiction.

For invertible bilateral shifts T we may consider the rational algebra A_R generated by the operator (that is, the weak operator closure of the rational functions in T). From Theorem 12 we know that this algebra coincides with $L^\infty(\beta)$. Let us call an operator *rationally strictly cyclic* is the algebra A_R is strictly cyclic. We omit the proof of the following analogue of Proposition 31.

PROPOSITION 36. *Let T be an invertible injective bilateral shift, represented as M_z on $L^2(\beta)$. Then T is rationally strictly cyclic if and only if $L^2(\beta) = L^\infty(\beta)$. In this case we have*

(1) $c\|f\|_\infty \leq \|f\|_2 \leq \|f\|_\infty \quad (f \in L^2(\beta))$,

(2) $r_3^- = r(T^{-1})^{-1}, \ r_2^+ = r(T)$,

(3) $s_n(f) \longrightarrow s(f)$ *in operator norm* $(f \in L^\infty(\beta))$,

(4) $\Sigma\, r(T)^{2n}/\beta(n)^2 < \infty, \ \Sigma\, r(T^{-1})^{-2n}/\beta(n)^2 < \infty$,

(5) $\Sigma\, |\hat{f}(n)| r(T)^n \leq c_1 \|f\|_2, \ \Sigma\, |\hat{f}(n)| r(T^{-1})^{-n} \leq c_2 \|f\|_2 \quad (f \in L^2(\beta))$.

The analogues of the two corollaries to Proposition 31 are valid, where now $\Delta(T)^-$ denotes the closed spectral annulus $r(T^{-1})^{-1} \leq |z| \leq r(T)$.

The analogue of Proposition 32 is also valid, with the change that now we have $-\infty < n, k, m < \infty$, in (59), (60), (61). The analogues of the first three corollaries are valid. In the third corollary the condition on $\{a_n\}$ is that (65) must hold for all n, m $(-\infty < n, m < \infty)$, where $\alpha(-n) = (a_{-1} \cdots a_{-n})^{-1}$ $(n > 0)$. In particular this will be the case if $a_0 \geq a_1 \geq \cdots \geq 1$ and if $\alpha(-n) = \alpha(n)$ for all n.

An example of an invertible, rationally strictly cyclic bilateral shift is obtained by choosing $\beta(n) = |n| + 1, \ -\infty < n < \infty$.

10. Invariant subspaces. A (closed) subspace M is said to be *invariant* for an operator T if $TM \subset M$. It is said to be *hyperinvariant* for T if it is invariant for each operator that commutes with T. It is an open question whether each operator on a Hilbert space has a proper invariant (or hyperinvariant) subspace. Of course, if the algebra generated by T coincides with the commutant, then every invariant subspace is hyperinvariant. In particular this is the case for injective unilateral shifts (see the corollary to Theorem 12 in §8).

The set of all the invariant subspaces of an operator T can be partially ordered by inclusion. It then becomes a complete lattice (a general reference on lattice theory is [**14**]; for lattices of invariant subspaces see [**80**], [**80**a], [**80**b], [**80**c], [**77**a]).

When we say that two operators have isomorphic lattices of invariant subspaces there are two things that can be meant. First, they are isomorphic as abstract lattices. Second, they are isomorphic as lattices of subspaces of Hilbert space; that is, there is a bounded invertible operator from one Hilbert space onto the other that maps the first lattice onto the second. In this case the lattices are said to be *spatially isomorphic*.

There are very few operators for which the lattice of all invariant subspaces is known. One such operator is the unweighted unilateral shift (see Chapter 7 of [**46**], or §6 of the article by Sarason in this volume). Its lattice will be called the Beurling lattice.

Notation. Lat (T) denotes the lattice of invariant subspaces of T.

Question 16. Let T be an injective unilateral shift, represented as M_z on $H^2(\beta)$. Suppose that Lat (T) is isomorphic, as an abstract lattice, to the Beurling lattice. What can be said about T?

If T is similar to the unweighted shift or, more generally, if T is similar to some nonzero scalar multiple of the unweighted shift then the lattices are isomorphic, in fact, spatially isomorphic. Are there any other cases? In particular, suppose $\beta(n) = (n+1)^{1/2}$ (the Dirichlet space), or $\beta(n) = (n+1)^{-1/2}$ (the Bergman space)? In these two cases it seems unlikely that the lattices will be isomorphic to the Beurling lattice. If Question 8 of §7 is answered in the negative for the Bergman space, then there are two invariant subspaces whose infimum (that is, whose intersection) is $\{0\}$. In the Beurling lattice the infimum of two nonzero elements is always nonzero, and therefore the lattices could not be isomorphic. For partial information on zero sets in the Dirichlet and Bergman spaces, see [**87**]. *Added in proof.* See the note at the end of this article.

If T is a strictly cyclic injective unilateral shift, then it follows from Corollary 2 to Proposition 31 in §9 that each invariant subspace lies in a maximal proper invariant subspace. The Beurling lattice does not have this property and so the lattices cannot be isomorphic.

Of special interest are the *quasi-analytic* shifts. These shifts have the following properties:

(i) $w_n \searrow 1$.

(ii) The shift is strictly cyclic.

(iii) If $f \in H^2(\beta)$ then each derivative of f has an absolutely convergent power series on $|z| = 1$, and hence is continuous on the closed unit disc. Thus we may speak of the order of a zero of f at a boundary point of the disc.

(iv) If $f \in H^2(\beta)$ and if f has infinitely many distinct zeros in the closed unit disc, or if f has a zero of infinite order at any point in the closed disc, then $f = 0$.

An example of such a shift is obtained by choosing $\beta(n) = \exp(n^{1/2})$. (i) and (iii) are obvious. (iv) is a theorem of Carleson (see [**15**, Theorem 3]). To establish (ii) we apply Corollary 1 to Proposition 32. Let $\alpha = 2 - 2^{1/2}$. We claim that

$$\frac{\exp[(k+m)^{1/2}]}{\exp[k^{1/2}]\exp[m^{1/2}]} \leqslant \exp[-\alpha k^{1/2}] + \exp[-\alpha m^{1/2}].$$

If $0 \leqslant k \leqslant m$ it will be sufficient to show that

$$\exp[(m+k)^{1/2}] \leqslant \exp[(1-\alpha)k^{1/2} + m^{1/2}];$$

that is, $(m+k)^{1/2} \leqslant (1-\alpha)k^{1/2} + m^{1/2}$. This is elementary.

Let T be a quasi-analytic unilateral shift, represented as M_z on $H^2(\beta)$. We obtain a class of invariant subspaces for T as follows. Let $z_1, \cdots, z_n$ be distinct points in the closed unit disc, and let $m_1, \cdots, m_n$ be positive integers. Let $I(z_1, \cdots, z_n; m_1, \cdots, m_n)$ denote the set of all those functions in $H^2(\beta)$ that have zeros at all the points z_i to at least the multiplicities m_i.

Question 17. If T is a quasi-analytic unilateral shift, are these the only invariant subspaces?

This would follow if it were known that T, restricted to any one of the invariant subspaces $I(z_1, \cdots, z_n; m_1, \cdots, m_n)$, is still strictly cyclic. (See the proof of Proposition 38 in this section.)

Let L be a family of subspaces of Hilbert space.

Notation. Alg L denotes the set of all those operators T that leave each subspace in L invariant, i.e., $L \subset \operatorname{Lat}(T)$. Alg L is a subalgebra of $L(H)$, and is closed in the weak operator topology.

DEFINITION. An operator T is said to be *reflexive* if $\operatorname{Alg}(\operatorname{Lat} T) = \mathsf{A}_T$, where A_T denotes the (weakly closed) algebra generated by T.

PROPOSTION 37. *If T is an injective unilateral shift and if T^* has a nonzero eigenvalue, then T is reflexive.*

PROOF. Let T be represented as M_z on $H^2(\beta)$. Since $\mathsf{A}_T = H^\infty(\beta)$ (Theorem 12 in §8) it will be sufficient to show that each operator S that leaves invariant all of the invariant subspaces of T must commute with T.

Since T^* has a nonzero eigenvalue, $r_2(T) > 0$ (see Theorem 10 in §6 and Theorem 8 in §5). Hence all the power series in $H^2(\beta)$ are convergent in the disc $\Delta_2(T) = \{z: |z| < r_2(T)\}$.

Since S leaves invariant all the invariant subspaces of M_z, S^* must do the same for M_z^*. Since the kernel functions k_w $(w \in \Delta_2(T))$ are simple eigenvectors for M_z^*, they must also be eigenvectors for S^*; let $(\phi(w))^-$ denote the corresponding eigenvalues $S^*k_w = (\phi(w))^- k_w$. We claim that S coincides with the operator of multiplication by ϕ on $H^2(\beta)$ (we view $H^2(\beta)$ as a space of analytic functions in $\Delta_2(T)$). Indeed, if $w \in \Delta_2(T)$ and if $f \in H^2(\beta)$ then

$$(Sf)(w) = (Sf, k_w) = (f, S^*k_w) = \phi(w)f(w).$$

In particular, taking $f = 1$ we see that $\phi = S1 \in H^2(\beta)$. Thus ϕ is representable by a power series, and $\phi H^2(\beta) \subset H^2(\beta)$; that is, $\phi \in H^\infty(\beta)$. This completes the proof.

Not all injective unilateral shifts are reflexive. We shall see some examples when we study unicellular shifts (see Proposition 39).

Question 18. Which shifts are reflexive?

An operator is said to be *unicellular* if its lattice of invariant subspaces is linearly ordered by inclusion. Every unilateral shift leaves invariant the subspaces E_n spanned by the basis vectors $\{e_n, e_{n+1}, \cdots\}$. These subspaces are linearly ordered, and a little consideration shows that if a unilateral shift is unicellular then it cannot have any other invariant subspaces (except $\{0\}$).

Recall that an operator is said to be *quasi-nilpotent* if 0 is the only point in its spectrum (that is, $r(T) = 0$).

PROPOSITION 38. *If T is a quasi-nilpotent, strongly strictly cyclic unilateral shift, then T is unicellular.*

PROOF. By Corollary 3 to Proposition 31 in §9 we see that f is a cyclic vector if and only if $\hat{f}(0) \neq 0$. By considering $T|E_n$ and applying the same reasoning we see that if $\hat{f}(0) = \cdots = \hat{f}(n-1) = 0$, but $\hat{f}(n) \neq 0$, then the cyclic subspace generated by f is precisely E_n. The result follows easily from this.

In the next three corollaries T is an injective unilateral shift with weight sequence $\{w_n\}$.

COROLLARY 1. *If* $\{w_n\}$ *is decreasing and in* l^p *for some* $p < \infty$, *then* T *is unicellular.*

PROOF. See Corollary 2 to Proposition 33.

COROLLARY 2. *If* $(n+1)w_n/(n+2)$ *is eventually decreasing and tends to zero, then* T *is unicellular.*

PROOF. See Proposition 34.

COROLLARY 3. *If* $w_n = 1/\log(n+2)$, *or if* $w_n = 1/\log\log(n+3)$, *then* T *is unicellular.*

PROOF. This follows from the preceding corollary.

The same holds for higher order iterated logarithms.

PROPOSITION 39. *A unicellular unilateral shift is never reflexive.*

PROOF. Let us write our operators as matrices with respect to the standard orthonormal basis $\{e_n\}$. If T is our shift then any commuting operator A is a formal power series in T, $A = \Sigma\, a_n T^n$ (see Theorem 3 in §4). Hence the entries in the diagonal immediately below the main diagonal are $a_1 w_0$, $a_1 w_1, a_1 w_2, \cdots$. In other words, $(Ae_n, e_{n+1}) = a_1 w_n$ $(n = 0, 1, \cdots)$. Thus if T_1 is another unilateral shift, whose weight sequence is not a scalar multiple of $\{w_n\}$, then T_1 does not commute with T and hence does not lie in the algebra generated by T. But T_1 leaves each subspace E_n invariant. This completes the proof.

Question 19. If $w_n \searrow 0$, is T unicellular?

Even if we assume in addition that the sequence $\{w_n\}$ is convex the answer is not known. (There are convex sequences decreasing to 0, to which Corollary 2 does not apply.)

There are several possible variations on Question 19. For example, if $w_n \searrow 0$ and if T is strictly cyclic, is T unicellular? Or again, if T is quasinilpotent and strictly cyclic, is it unicellular? (Recall Question 14: Does strict cyclicity imply strong strict cyclicity?)

We now give two examples. The first is of a unicellular shift T whose weight sequence does not tend to 0.

EXAMPLE 1. Let $w_{2k} = 1$, $w_{2k+1} = 1/(k+1)$ $(k = 0, 1, \cdots)$. Let $u_n = 1/(n+1)$, and $\alpha(n) = u_0 \cdots u_{n-1}$. Let $\beta(k, n) = \beta(n)/\beta(k)\beta(n-k)$, $0 \leqslant k \leqslant n$, and define $\alpha(k, n)$ similarly. A little checking shows that

$$\sum_{k=0}^{2n+1} \beta(k, 2n+1)^2 = 2 \sum_{k=0}^{n} \alpha(k, n)^2 < c,$$

$$\sum_{k=0}^{2n} \beta(k, 2n+1)^2 < 3 \sum_{k=0}^{n} \alpha(k, n)^2 < c \qquad (n = 0, 1, \cdots).$$

It follows from Proposition 32 (61) that T is strictly cyclic. Similarly one sees that T is strongly strictly cyclic. Also, T is quasi-nilpotent. Hence, by Proposition 38, T is unicellular.

Question 20. If the unilateral shift T is unicellular, must it be quasi-nilpotent? strictly cyclic?

C. Foiaş and J. Williams **[26]** have given an example in a different context of a unicellular operator with more than one point in its spectrum.

We now give an example of a nonunicellular shift T with weights $w_n \to 0$ (not monotonically). We require the following lemma, whose proof we omit.

LEMMA. *Let S and A be operators on Hilbert space with $\|A\| < 1$, and S an isometry that is not surjective. Then $S + A$ has closed range and is not surjective.*

EXAMPLE 2. Let $n_k = \frac{1}{2}(k+1)(k+2) - 1$, $k = 0, 1, \cdots$. Let $\{u_n\}$ be given, with $u_n \downarrow$, and $u_n > 0$ $(n = 0, 1, \cdots)$. Let T be the weighted shift with weight sequence $\{w_n\}$, where

$$w_j = u_j \quad (j \notin \{n_k\}), \qquad w_{n_k} = v_k \quad (k = 0, 1, \cdots).$$

The sequence $\{v_k\}$ must satisfy certain conditions that will be specified later. Let $x = \{x(n)\} \in l^2$ be given, with $x(0) = 1$, $x(n_k + 1) = a_k > 0$ $(k = 0, 1, \cdots)$, and $x(n) = 0$ otherwise. Let $\{a_k\}$ satisfy

$$(66) \qquad \sum_{0}^{\infty} a_k^2 < \frac{1}{2}, \quad \sum_{i+1}^{\infty} \left(\frac{a_k}{a_i}\right)^2 < \frac{1}{2^{i+2}} \qquad (i = 0, 1, \cdots).$$

We shall show that x is not a cyclic vector for T. Since x is not in any of the subspaces E_n, there must be additional invariant subspaces and so T is not unicellular. The following table shows the vectors $x, Tx, T^2x, T^3x, \cdots$ as row vectors in l^2.

	n_0		n_1			n_2				n_3
	0	1	2	3	4	5	6	7	8	9
x:	$\boxed{1}$	a_0	0	a_1	0	0	a_2	0	0	0
Tx:	0	v_0	$\boxed{a_0u_1}$	0	a_1u_3	0	0	a_2u_6	0	0
T^2x:	0	0	v_0u_1	$a_0v_1u_1$	0	$\boxed{a_1u_3u_4}$	0	0	$a_2u_6u_7$	0
T^3x:	0	0	0	$v_0v_1u_1$	$a_0v_1u_1u_2$	0	$a_1v_2u_3u_4$	0	0	$\boxed{a_2u_6u_7u_8}$

The boxes are placed around what will turn out to be the largest entry in each row. In general the dominant term for T^k occurs in position n_k. Let α_k be the reciprocal of the coefficient of this dominant term.

Define an isometry S on l^2 by $S_{e_k} = e_{1+n_k}$. Define an operator A by

$$Ae_k = \alpha_k T^k x - Se_k.$$

If we knew that $\|A\| < 1$, then we would have $T^k x \in \text{range}(S + A)$, $k = 0, 1, \cdots$. By the lemma this range is a proper closed subspace, and so x is not a cyclic vector.

Thus to complete the proof it will be sufficient to show that the Hilbert-Schmidt norm of A is less than 1. This amounts to the following. In our table divide each row by the dominant term, so that in the new table the entry 1 will occur in each box. Delete these from the table. Then show that the sum of the squares of all the remaining entries is less than 1.

The sum of squares in the first row is less than ½, by the choice of $\{a_k\}$. In the second row the entries to be considered are

$$v_0/a_0u_1, \quad a_1u_3/a_0u_1, \quad a_2u_6/a_0u_1, \cdots.$$

We may replace each u_k/u_1 by 1. From (66) we have $\Sigma\,(a_k/a_0)^2 < 1/8$. Hence if we restrict v_0 so that $(v_0/a_1u_1)^2 < 1/8$, then the contribution from the second row will be less than 1/4. We shall place one more restriction on v_0.

In the next row we use the monotonicity of the $\{u_n\}$ and (66) to handle the terms beyond the dominant one. They contribute at most 1/16 to the sum of squares. To handle the two terms that precede the dominant term we require

$$\left(\frac{v_0u_1}{a_1u_3u_4}\right)^2 + \left(\frac{a_0v_1u_1}{a_1u_3u_4}\right)^2 < \frac{1}{16}.$$

We now choose v_0 so that $(v_0/a_0u_1)^2 < 1/8$, and $(v_0u_1/a_1u_3u_4)^2 < 1/32$. Also, the first restriction that v_1 must satisfy is that $(a_0v_1u_1/a_1u_3u_4)^2 < 1/32$. Note that in all future rows wherever v_0 occurs it is multiplied by v_1. We continue in this manner, placing a finite number of restrictions on v_1. After a certain point v_1 only occurs multiplied by v_2, etc.

P. Halmos asked (see [**35**, Question 2]) whether the operator obtained by restricting a unilateral weighted shift to an invariant subspace is always similar to a weighted shift. R. Gellar [**30**] gave an example to show that this is not always true. The idea of the example is the following. Let $\beta(n) = n + 1$ (thus $w_n = (n + 2)/(n + 1)$). Then each boundary point $|z| = 1$ of the spectral disc is a simple eigenvalue for T^*, and thus is a bounded point evaluation on $H^2(\beta)$. Let $\mathcal{M}$ be the subspace of $H^2(\beta)$ consisting of those functions that vanish at $z = 1$. This is an invariant subspace for M_z. Let $A = M_z | \mathcal{M}$. It can be shown that A^* does not have $w = 1$ as an eigenvalue. It does have all other w of modulus 1 as eigenvalues, however. (The corresponding eigenvectors are obtained by projecting into $\mathcal{M}$ the eigenvectors for M_z^*.) Thus the eigenvalues fail to have circular symmetry. Hence A cannot be similar to a weighted shift.

We briefly indicate the proof that $z = 1$ is not an eigenvalue of A^*. Assume $A^*f_0 = f_0$ for some $f_0 \neq 0$ in $\mathcal{M}$. Then $(M_z^* f_0, g) = (f_0, g)$, or $(f_0, (z-1)g) = 0$, for all $g \in \mathcal{M}$. In particular $(f_0, (z-1)^2 h) = 0$ for all h in $H^2(\beta)$. It can be shown that the set $(z-1)^2 H^2(\beta)$ is dense in $\mathcal{M}$. Hence f_0 is orthogonal to $\mathcal{M}$, which is a contradiction.

One can still ask: When is the restriction of a shift to an invariant subspace similar to a shift? In particular is this always true for quasi-analytic shifts? for quasi-nilpotent shifts? for M_z on $H^2(\beta)$ when $\beta(n) = (n+1)^{1/2}$? when $\beta(n) = (n+1)^{-1/2}$? Perhaps one should first raise the following problem.

Question 21. If T is an injective unilateral shift and $\mathcal{M}$ an invariant subspace, does $(T\mathcal{M})^-$ always have codimension 1 in $\mathcal{M}$?

We now say a few words about bilateral shifts. Let E_n denote the (closed) span of $\{e_n, e_{n+1}, \cdots\}$, $-\infty < n < \infty$. Every bilateral shift T leaves these subspaces invariant. It is easily seen that T is unicellular if and only if these are the only invariant subspaces.

Question 22. Are there any unicellular bilateral shifts?

It is not known whether there is any operator A on Hilbert space for which $\mathrm{Lat}(A)$ is isomorphic to the integers. Banach algebra methods do not seem to help here.

Question 23. Does every bilateral shift have a hyperinvariant subspace?

Of courst this question is only of interest for invertible shifts T such that neither T nor T^* has any eigenvectors. One such T is the unweighted bilateral shift $(w_n = 1)$. Here hyperinvariant subspaces are known to exist.

One could define quasi-analytic bilateral shifts. Again it seems reasonable to conjecture that the only invariant subspaces would be those determined by zeros, with multiplicities.

11. Cyclic vectors. Recall that a vector f is called a cyclic vector for an operator T if the smallest closed invariant subspace containing f is all of H. In other words, $\mathcal{A}_T f$ is dense in H, where $\mathcal{A}_T$ denotes the closure, in the weak operator topology, of the polynomials in T. Equivalently, f is a cyclic vector if it lies in no invariant subspace. Thus the set of cyclic vectors is the complement of the union of all the invariant subspaces.

PROPOSITION 40. *If T is any operator on any separable Banach space X, then the set of all cyclic vectors for T is a Borel set of type G_δ.*

PROOF. Let $\{O_n\}$ be a countable basis for the open subsets of X. Fix n. The set of vectors f such that $\mathsf{A}_T f$ meets O_n is an open set. Now let n vary. The set of cyclic vectors is the intersection of these open sets. This completes the proof.

For our purposes we distinguish three types of G_δ sets: open, closed, neither. Are there operators on Hilbert space for which the set of cyclic vectors is an open set? a closed set? neither? The set of cyclic vectors is never a closed set (if it is nonempty) for a trivial reason. Namely, if f is cyclic so are the vectors f/n, which converge to 0. But 0 is not cyclic. Thus we shall consider $H\backslash\{0\}$ instead of H, and "open," "closed," and "neither" will all be with respect to $H\backslash\{0\}$.

For the case of an open set the answer is affirmative. For example, consider the finite dimensional shift $Te_k = e_{k+1}$, $1 \leqslant k < n$, $Te_n = 0$. The set of cyclic vectors is precisely the set of vectors with first coordinate different from 0. This example is a special case of the following. Let T be a strictly cyclic operator (on any Banach space X). Then X becomes a Banach algebra (see the first part of §9) and so a vector is a cyclic vector if and only if it lies in no maximal ideal, that is, if and only if its Gelfand transform never vanishes on the maximal ideal space. The set of such f is an open set.

The cyclic vectors of the unweighted unilateral shift $(w_n = 1)$ are known; they are precisely the outer functions in H^2. These functions are not closed in $H^2\backslash\{0\}$. For example, if ϕ is a singular inner function, if $r_n \nearrow 1$, if $\phi_n(z) = \phi(r_n z)$, then each ϕ_n is an outer function but $\phi_n \longrightarrow \phi$. On the other hand the set of cyclic vectors is not an open set. For example, the function 1 is a cyclic vector, but it is the limit of the inner functions $(z - r_n)/(1 - r_n z)$, if $r_n \longrightarrow 1$.

Question 24. Is there an operator on Hilbert space for which the set of cyclic vectors is a nonempty closed subset of $H\backslash\{0\}$.

Of course an operator with no proper invariant subspaces would be an example. Excluding this we have the equivalent question: Can the set of noncyclic vectors be a proper open subset of $H\backslash\{0\}$? One might hope for an example as a subnormal shift of norm 1 (see §7). Here $H^2(\beta)$ consists of analytic functions in the open unit disc. The only obvious sufficient condition for noncyclicity of f is that f have a zero in the open unit disc. If β can be chosen so that this condition is also necessary, then the set of noncyclic vectors would be an open subset of $H^2(\beta)\backslash\{0\}$. Indeed, by Hurwitz' theorem in complex function theory, if $f(z_0) = 0$, if f is not constant, and if $f_n \longrightarrow f$ uniformly on compact subsets, then for large n each f_n has a zero near z_0. This brings us very close to Beurling's original point of view in studying the shift operator (see [**12**], and also [**13**]).

Question 24′. Does there exist β with $r_2(M_z) = \|M_z\| = 1$, and such that a vector $f \in H^2(\beta)$ is cyclic if and only if it has no zeros in the open unit disc?

We now consider the Bergman space, $\beta(n) = (n+1)^{-1/2}$. Here the singular inner function $\phi(z) = \exp[(z+1)/(z-1)]$ has no zeros in the open unit disc, and is in $H^2(\beta)$, but is not a cyclic vector. This was first proved by M. V. Keldyš [**49**]. We give the proof here briefly.

Assume that ϕ were cyclic: $p_n\phi \longrightarrow 1$ in $H^2(\beta)$, where p_n are polynomials. We have $|f(w)| \leqslant \|f\|\,\|k_w\|$ for f in $H^2(\beta)$, where $k_w(z) = 1/(1-\bar{w}z)^2$ is the Bergman reproducing kernel (see §6, equations (35)–(37)). Hence

$$|f(z)| \leqslant \|f\|/(1-|z|^2), \qquad f \in H^2(\beta). \tag{67}$$

Applying this to the bounded sequence $\{p_n\phi\}$ we obtain $|p_n(z)\phi(z)| \leqslant c/(1-|z|^2)$.

Let $g(z) = 1/(1-z)^2$, and let D denote the disc $|z - \tfrac{1}{2}| < \tfrac{1}{2}$. Note that ϕ is bounded away from 0 on ∂D. Hence

$$|p_n(z)| < c\,|g(z)|, \qquad z \in \partial D \neq 1,$$

or, p_n/g is bounded by c, independent of n, on ∂D, except possibly for one point. (Note that the constant c is not the same at each occurrence.) But g is an outer function and so p_n/g is in the class N^+. By a form of the maximum principle (see [**24**, Chapter 2, §5, Theorem 2.11]), we have $|p_n/g| \leqslant c$ in D, for all n. But $p_n(z) \longrightarrow 1/\phi(z)$ for all $|z| < 1$, and so $1 \leqslant c\,|\phi g|$ in D. But this is absurd, since ϕ tends to 0 exponentially on the radius $0 < x < 1$.

On the other hand H. S. Shapiro has shown ([**84**, Theorem 2], and [**86**, Theorem 1]) that if ϕ is a singular inner function for which $|\phi(z)| > c(1-|z|)^\epsilon$, for some c, ϵ and for all $|z| < 1$, then ϕ is a cyclic vector for the Bergman space. Incidentally, he shows that the lower bound for ϕ is equivalent to the associated singular measure being continuous, with modulus of continuity $\leqslant c_1 t \log(1/t)$.

More generally Shapiro shows [**86**, Theorem A] that if $f \in H^2(\beta^*)$ where $\beta^*(n) = (n+1)^{-\alpha}$, for some $\alpha < 1/2$, and if, for some c, $\epsilon > 0$,

$$|f(z)| \geqslant c(1 - |z|)^\epsilon \qquad (|z| < 1), \tag{68}$$

then f is a cyclic vector for the Bergman space. This suggests the following problem.

Question 25. If f is in the Bergman space and satisfies (68), is f a cyclic vector?

This is known to be true if some additional condition is imposed. One such is the result of Shapiro mentioned above, where f is required to lie in a smaller Hilbert space. Another such additional condition is given in Theorem 2 of [**1**], where it is shown that if, in additon to (68), the range of f omits a set of positive logarithmic capacity, then f is a cyclic vector.

Still another condition is given by L. I. Hedberg, who proved the following result. We include an indication of the proof, since it is not being published elsewhere.

PROPOSITION 41 (HEDBERG). *If f is in the Bergman space and if f is the derivative of a univalent function, then f is a cyclic vector.*

Note. Condition (68) is automatically satisfied when f is the derivative of a univalent function. This is the "distortion theorem" of conformal mapping (see [**65**, Theorem 1.7, p. 17]).

Notation. If G is a plane domain of finite area, then $L^2_a(G)$ denotes the space of square integrable analytic functions in G (with respect to area measure). This is known to be a closed subspace of $L^2(G)$. See [**9**] for a study of this space. Recall also that $H^\infty(G)$ denotes the space of bounded analytic functions in G. H^∞ denotes the space of bounded analytic functions in the unit disc.

PROOF. We must show that $H^\infty f$ is dense in the Bergman space. Since f is the derivative of a univalent function, let G denote the range of this univalent function. Then G is a simply connected domain of finite area. By a change of variables in the area integral, our problem is equivalent to the following result (which was the form in which Hedberg proved the result).

THEOREM. *$H^\infty(G)$ is dense in $L^2_a(G)$.*

PROOF. We must show that if $g \in L^2(G)$, and if $\int gh = 0$ for all $h \in H^\infty(G)$, then $\int gf = 0$ for all $f \in L^2_a(G)$. The integrals are taken with respect to area measure on G.

Let $\delta(z) = \operatorname{dist}(z, \partial G)$. Let $D_z(r)$ denote the open disc of center z and radius r. By Lemma 20.2, p. 383 in **[83]**, there exists $c > 0$ such that for each given point $z \in G$ there exists $h_z \in H^\infty(G)$ (in fact, h_z is bounded and holomorphic outside of $D_z(2\delta(z)) \cap \partial G$) satisfying

(i) $|h_z(w)| \leq c/\delta(z) \quad (w \in G)$,

(ii) $|1/(w-z) - h_z(w)| \leq c[\delta(z)]^2/|w-z|^3 \quad (w \in G)$.

Next, let n be a given positive integer. Choose a real-valued function ϕ_n, continuously differentiable in the whole plane, such that $0 \leq \phi_n \leq 1$, and

$$\phi_n(n) = 1 \quad \text{if } \delta(z) \geq 2/n \quad (z \in G),$$

$$\phi_n(z) = 0 \quad \text{if } \delta(z) \leq 1/n, \quad \text{or if } z \notin G,$$

$$|\operatorname{grad} \phi_n| \leq 2n.$$

It will be sufficient to prove that $\int fg\phi_n \longrightarrow 0$, if $g \in L^2(G)$ and $\int gh = 0$ for all $h \in H^\infty(G)$.

We require some further notations and facts. Here μ will denote a measure of compact support in the plane.

$$\bar{\partial} = \frac{\partial}{\partial \bar{z}} = \frac{1}{2}\left(\frac{\partial}{\partial x} - \frac{1}{i}\frac{\partial}{\partial y}\right).$$

Thus the Cauchy-Riemann equations are equivalent to $\bar{\partial} h = 0$.

$$\hat{\mu}(z) = \int \frac{1}{w-z}\, d\mu(w)$$

is called the Cauchy transform of μ. We have $\mu = -\pi^{-1}\bar{\partial}\hat{\mu}$ in the distributional sense.

Integrating by parts we have

$$\int f\phi_n g = -\frac{1}{\pi}\int f\phi_n \bar{\partial}\hat{g} = \frac{1}{\pi}\int \bar{\partial}(f\phi_n)\hat{g} = \frac{1}{\pi}\int f\hat{g}\,\bar{\partial}\phi_n.$$

Thus, letting $G_n = \{z\colon\ 1 < n\delta(z) < 2\}$,

$$\left|\int fg\phi_n\right| \leq cn\int_{G_n} |f\hat{g}| \leq cn\left(\int_{G_n} |f|^2\right)^{1/2}\left(\int_{G_n} |\hat{g}|^2\right)^{1/2} \tag{69}$$
$$\leq cn\|f\|\left(\int_{G_n} |\hat{g}|^2\right)^{1/2}.$$

Since g is orthogonal to $H^\infty(G)$ we have

$$\hat{g}(z) = \int \frac{g(w)}{w-z} = \int g(w)\left(\frac{1}{w-z} - h_z(w)\right).$$

Let $A_n = \{w\colon\ |w-z| \leq 1/n\}$, $B_n = \{w\colon\ |w-z| > 1/n\}$. Then

$$
(70)\qquad
\begin{aligned}
|\hat{g}(z)| &\leqslant \int_{A_n} |g(w)| \left(\frac{1}{|w-z|} + \frac{c}{\delta(z)} \right) + c\delta(z)^2 \int_{B_n} |g(w)| \frac{1}{|w-z|^3} \\
&\leqslant c\int_{A_n} \frac{|g(w)|}{|w-z|} + \frac{c}{n^2}\int_{B_n} \frac{|g(w)|}{|w-z|^3},
\end{aligned}
$$

where c is not necessarily the same constant at each occurrence.

Let

$$
g^*(z) = \sup_{a>0} \frac{1}{\pi a^2} \int_{|w-z|<a} |g(w)|
$$

be the Hardy-Littlewood maximal function. Let $\epsilon = 1/n$. We now estimate separately the two terms on the right side of (70). We begin with the second term.

$$
\begin{aligned}
\epsilon^2 \int_{|w-z|>\epsilon} \frac{|g(w)|}{|w-z|^3} &\leqslant \sum_0^\infty \epsilon^2 \left(\frac{1}{\epsilon 2^k}\right)^3 \int_{\epsilon 2^k<|w-z|<\epsilon 2^{k+1}} |g(w)| \\
&\leqslant \frac{1}{\epsilon}\sum_0^\infty 2^{-3k}(\pi\epsilon 2^{k+1})^2 \left(\frac{1}{\pi\epsilon 2^{k+1}}\right)^2 \int_{|w-z|\leqslant \epsilon 2^{k+1}} |g(w)| \\
&\leqslant c\epsilon \sum_0^\infty 2^{-k} g^*(z) = c\epsilon g^*(z).
\end{aligned}
$$

For the first term we have

$$
\begin{aligned}
\int_{|w-z|\leqslant\epsilon} \frac{|g(w)|}{|w-z|} &= \sum_0^\infty \int_{\epsilon/2^{k+1}<|w-z|\leqslant\epsilon/2^k} \frac{|g(w)|}{w-z} \\
&\leqslant \sum_0^\infty \left(\frac{\pi\epsilon}{2^k}\right)^2 \frac{2^{k+1}}{\epsilon}\left(\frac{2^k}{\pi\epsilon}\right)^2 \int_{|w-z|\leqslant\epsilon/2^k} |g(w)| \\
&\leqslant c\epsilon \sum 2^{-k} g^*(z) = c\epsilon g^*(z).
\end{aligned}
$$

Thus we finally have $|\hat{g}(z)| \leqslant (c/n) g^*(z)$. Hence from (69),

$$
\left|\int fg\phi_n\right| \leqslant cn\|f\| \frac{c}{n}\left(\int_{G_n} |g^*|^2\right)^{1/2}.
$$

By the Hardy-Littlewood maximal theorem $g^* \in L^2(G)$ and so the last term tends to 0 as $n \to \infty$, which completes the proof.

The following special case of Question 25 seems to be unknown.

Question 25′. If f and $1/f$ are both in the Bergman space, is f a cyclic vector?

Note that the lower bound condition (68) is automatically satisfied here because of (67).

We have seen that in the Bergman space some inner functions are cyclic vectors. This suggests the following question for the Dirichlet space ($\beta(n) = (n + 1)^{1/2}$).

Question 26. Is there an outer function in the Dirichlet space that is not a cyclic vector?

Probably such functions exist. The reverse implication is true: Every cyclic vector in the Dirichlet space is an outer function. Indeed, if $p_n f \longrightarrow 1$ in the Dirichlet norm then we also have convergence in H^2 and so f is cyclic in H^2. This is related to the problem raised in the second paragraph following the statement of Question 6 in §6, since if ϕ is bounded away from 0 in the open unit disc, then it is an outer function.

We turn now to bilateral shifts.

PROPOSITION 42. *Let T be an injective bilateral weighted shift, represented as M_z on $L^2(\beta)$.*

(a) *If T^* has an eigenvalue, then T has no cyclic vectors.*

(b) *If T has an eigenvalue, then T has a cyclic vector.*

PROOF. For the basic facts about eigenvectors of bilateral shifts see Theorem 9 in §5, especially (34). The eigenvalues are all simple.

(a) By circular symmetry the set of eigenvalues for T^* contains a circle about the origin. Without loss of generality we shall assume that this circle is the unit circle Γ. Let k_w be the corresponding eigenvectors: $T^*k_w = \bar{w}k_w$ ($w \in \Gamma$). Assume that the k_w satisfy the condition: $\hat{k}_w(0) = 1$.

If $g \in L^2(\beta)$ then the Laurent series g converges absolutely at w to the value $g(w) = (g, k_w)$ (see Theorem 10′(ii) at the end of §6). Hence $|g(w)| \leqslant \|g\| \, \|k_w\|$. A little checking shows that $\|k_w\|$ is constant for $|w| = 1$, and in fact, $g(w)$ is a continuous function of w. We regard the elements of $L^2(\beta)$ as functions defined on the point spectrum of T^*. Then the restriction map $g \mapsto g|\Gamma$ may be regarded as a bounded linear map of $L^2(\beta)$ into $C(\Gamma)$. This map has dense range since the Laurent polynomials map onto the trigonometric polynomials.

If T had a cyclic vector f, then $f|\Gamma$ would be cyclic for the operator of multiplication by z on $C(\Gamma)$. Hence f must have no zeros on Γ. But then the operator of multiplication by f on $C(\Gamma)$ would be invertible. Hence the image of a proper closed subspace would be a proper closed subspace. In particular, if $A(\Gamma)$ denotes the disc algebra (the closure of the polynomials in $C(\Gamma)$), then $fA(\Gamma)$ is a proper closed subspace of $C(\Gamma)$. Hence $f|\Gamma$ is not a cyclic vector in $C(\Gamma)$. This completes the proof of (a).

Note that the set $\{k_w\}$ is total. Indeed, if g is orthogonal to this set, then the Laurent series g converges absolutely to zero at each point of the unit circle, and hence is identically zero. Likewise any countable subset of $\{k_w\}$, corresponding to a dense subset of Γ is total.

(b) We may assume that Γ is contained in the point spectrum of T. Let $\{w_n\}$ be an enumeration of all the roots of unity; then $\{w_n\}$ is a dense subset of Γ. Fix m, n. Then $\bar{w}_m w_n$ is a primitive kth root of unity for some k, and $k = 0$ if and only if $m = n$. Hence, for sufficiently large N,

$$\frac{1}{N!}\sum_{j=0}^{N!} (\bar{w}_m w_n)^j = \delta_{mn}. \tag{71}$$

Let h_n denote an eigenvector of T corresponding to the eigenvalue w_n, with $\|h_n\| = 1$. Let $\{a_n\} \in l^1$ be given, with $|a_n| > 0$ for all n. Let $g = \Sigma\, a_n h_n$. Thus $T^j g = \Sigma\, a_n w_n^j h_n$. Fix m. For large N we have, from (71),

$$\frac{1}{N!}\sum_{j=1}^{N!} (\bar{w}_m T)^j g = a_m h_m + \sum_{n=M}^{\infty} a_n \frac{1}{N!}\sum_{j=1}^{N!} (\bar{w}_m w_n)^j h_n$$

where $m < M$, and $M \longrightarrow \infty$ as $N \longrightarrow \infty$. It follows that h_m is in the cyclic subspace spanned by g. But the set $\{h_m\}$ is total. This is proved just as we proved at the end of (a) that the set $\{k_w\}$ is total. Hence g is a cyclic vector, which completes the proof.

Question 27. If T is an injective bilateral shift, must T or T^* have a cyclic vector?

In the case of the unweighted bilateral shift T, neither T nor T^* has any eigenvectors, but they both have cyclic vectors, in fact, the same cyclic vectors. These are the vectors $f \in L^2(\Gamma)$ such that $|f| > 0$ almost everywhere, and $\int \log |f| = -\infty$.

12. Notes and comments. Weighted shift operators have occurred sporadically in the literature for a number of years as examples and counterexamples. The first systematic study was R. L. Kelley's thesis **[50]** which was, unfortunately, never published. The results in §§1 and 2 are due to him. He also discusses reducing subspaces of shifts. An injective unilateral shift has no reducing subspaces (see Corollary 2 to Theorem 3 in §4 of the present work). An injective bilateral shift with weight sequence $\{w_n\}$ has a nontrivial reducing subspace if and only if the sequence $\{|w_n|\}$ is periodic. This result was also obtained by N. K. Nikolskiĭ (see Theorem 4 in **[70]**).

It has long been folklore that a weighted shift operator could be viewed as "multiplication by z" on a Hilbert space of formal power series. This point of view was taken by R. Gellar in [27], [28], and [29], where he considered much more general spaces, and by N. K. Nikolskiĭ [71], and more recently by S. Grabiner [33].

The commutant of a weighted shift operator was first described, matricially, by R. L. Kelley [50, p. 5]. More precisely, he proved Proposition 5 of §2 of the present paper. The more useful description of the commutant (of a bilateral shift) as a space of formal Laurent series occurs in Gellar [28]. The notations $H^\infty(\beta)$ and $L^\infty(\beta)$ were suggested by Gellar [30], and by A. L. Shields.

The fact that the commutant of an operator on Hilbert space is isometric to a conjugate Banach space was pointed out to me by W. B. Arveson.

The material in §5 on the parts of the spectrum is mostly due to R. L. Kelley [50]. However, he did not identify the approximate point spectrum. This was done by W. Ridge [79]. Independently, N. K. Nikolskiĭ identified the eigenvalues of an injective bilateral shift (see [70, Theorem 5]).

In the special case when the weight sequence is monotone increasing the parts of the spectrum were identified earlier by J. G. Stampfli [91, Theorem 1].

The basic facts about analytic structure (§6), including most parts of Theorems 10 and 10′, are due to Gellar [28], [29]. Part (vii) of Theorem 10 is due to M. Rabindranathan. In carrying his proof over to invertible bilateral shifts (part (vii)(a) of Theorem 10′) we were led to Proposition 23 and to Question 7.

The study of subnormal weighted shifts was begun by J. Stampfli [91]. Proposition 25 was apparently found independently by C. Berger (see [35, p. 895]), D. Sarason (see [57]), and by R. Gellar and L. Wallen [32]. Proposition 27 is due to D. Herrero [41, Theorem 2]. Question 9 arose in conversation with M. Rabindranathan. The following problem seems to be open.

Question 28. Is every hyponormal shift the sum of a subnormal shift and an operator of the trace class?

The *trace class* is the class C^1 of Proposition 4 (§2).

The first general result on the algebra generated by a shift was obtained by Shields and Wallen [88], who proved Theorem 12 (§8) for unilateral shifts. Their method was different, and was based on a result of I. Schur (though the Fejer kernel also played a key role).

In his thesis [96, Proposition 2.9] T. R. Turner showed that if a bilateral shift is injective but not invertible, then the algebra generated by the shift coincides with the commutant. He also showed that if a unilateral shift has some zero weights then the commutant is nonabelian, and in this case the algebra coincides with the second commutant. Independently of the present work, Gellar [**30**, §2] and Gellar and Herrero [**31**, Theorem 1] have obtained Theorem 12 for all bilateral shifts (on more general spaces). They also say something about Fourier coefficient multipliers (see §3 of [**31**]). The presentation given in §8 is suggested by the theory of Fourier series. A similar point of view is taken by K. de Leeuw [**20**] when studying operators on the space of continuous functions on the circle group.

Strictly cyclic operators and algebras were first defined by A. L. Lambert in his thesis [**58**] (see also [**59**] and [**60**]). He was motivated in part by a paper of Eric Nordgren [**74**]. It had been recognized earlier that for some weighted shifts one had $H^2(\beta) = H^\infty(\beta)$ (see, for example, Theorem 22, p. 78 in Gellar's paper [**29**]). Some properties of such spaces were discussed in a seminar at the University of Michigan in early 1967. But Lambert was the first to formulate the important concept of a strictly cyclic operator.

Proposition 30 was proved originally by Lambert for abelian strictly cyclic algebras by a more complicated proof. The statement and proof of the text (§9) were shown to me by E. Azoff, and were found independently by him and by Lambert, and by H. Radjavi and P. Rosenthal [**77a**, Corollary 9.9, p. 172].

The corollaries to Proposition 32 occur, in modified form, in Lambert's paper [**59**]. He also shows (p. 336) that the unilateral shift with weight sequence $w_n = 1/\log(n+2)$ is strictly cyclic. Proposition 32 itself is due to E. Kerlin and A. Lambert [**51**].

Some sufficient conditions more restrictive than those of Proposition 32 are given by Gellar [**29**, Theorems 22, 23]. His theorem applies to more general Banach spaces of formal power series.

Gellar has also considered a question equivalent to the question of when a shift operator is strongly strictly cyclic (see [**29**, Theorem 24]). He gives sufficient conditions for this to occur (again, for quite general Banach spaces of formal power series) and, in particular, he obtains our Corollary 2 to Proposition 33 (see [**29**, the remark following Note 3, on the middle of p. 80]).

Example 3 was found jointly with M. Rabindranathan.

For bilateral shifts Gellar considered the question of when $L^2(\beta) = L^\infty(\beta)$ (see [**29**, §10, pp. 74–76]). As always he studied more general spaces.

In connection with lattices of invariant subspaces, and with reflexive operators, see Problems 9 and 10 in [**35**], also J. Deddens [**19**], and K. J. Harrison [**37**]. We mention the following problem, due to P. Rosenthal and D. Sarason.

Question 29. If A, B are commuting operators on Hilbert space with $\operatorname{Lat} A \subset \operatorname{Lat} B$, must B belong to the (weakly closed) algebra generated by A?

More complete conditions for quasi-analyticity of classes of functions analytic in the unit disc may be found in B. I. Korenblium [53]. Information about the zeros of some classes of analytic functions in the disc, with some smoothness on the boundary, may be found in B. A. Taylor and D. L. Williams [94], J. D. Nelson [67], B. I. Korenblium [54], and L. Carleson [15].

Some results about invariant subspaces in spaces of analytic functions with some smoothness on the boundary of the unit disc may be found in Taylor and Williams [93], and in Korenblium [55], [56].

A description of invariant subspaces of finite codimension is in Gellar [30].

There are a number of sufficient conditions for a unilateral shift to be unicellular. See, for example, N. K. Nikolskiĭ [69] where it is shown (Theorem 2), that if the weight sequence w_n is monotone decreasing, and in l^p for some $p < \infty$ (equivalently, $w_n \leq c/n^\epsilon$ for some $\epsilon > 0$), then the operator is unicellular. This was also found by S. Parrott and by Shields. More refined conditions are given in Nikolskiĭ [70, Theorem 1] and [71], and in K. J. Harrison [36]. Helson [40] gives a Banach algebra proof of Nikolskiĭ's basic result. In [71, Theorem 4], Nikolskiĭ indicates that the shift with weight sequence

$$1/n^{\epsilon_0}(\log n)^{\epsilon_1}(\log\log n)^{\epsilon_2} \cdots (\log\cdots\log n)^{\epsilon_k} \qquad (n \text{ large})$$

is unicellular (ϵ_i are nonnegative and at least one is positive). Proposition 38 of the present text first occurs in Gellar [29, Theorem 24]; this result of Gellar's seems to have been generally overlooked.

For more general Banach algebra versions of unicellularity see D. Beckles [6], [7], S. Grabiner [33], and D. Herrero [42].

Example 2 of §10 (a nonunicellular unilateral shift with weight sequence tending to 0) is modified from Nikolskiĭ [70, Theorem 3].

Reflexive bilateral shifts are studied by Herrero and Lambert in [44]. The idempotent operators (not necessarily selfadjoint) that commute with a bilateral shift are studied by Gellar in §2 of [30]. Finally, hyperinvariant subspaces of bilateral weighted shifts are studied by Gellar and Herrero in [31]. They also give examples of bilateral shifts with the property that every operator in the commutant is one-to-one and has dense range.

Proposition 40 of §11 is due to H. S. Shapiro (unpublished). Nikolskiĭ [73] has shown that for a wide class of weighted sup-norm Banach spaces of analytic functions in the unit disc there are always nonvanishing functions that are not cyclic vectors. This suggests that the answer to Question 24′ is probably negative.

H. S. Shapiro has observed to the author that the Keldyš example,

$\phi(z) = \exp[(z+1)/(z-1)]$, of a nonvanishing function that is not cyclic in the Bergman space can be strengthened. More precisely, we have the following result. Recall that the "Bergman space" is $H^2(\beta)$, $\beta(n) = (n+1)^{-1/2}$.

PROPOSITION 43. *Let* ϕ *be the Bergman space, with*

$$\phi(r) \leqslant \exp[a/(1-r)], \qquad 0 < r < 1, \tag{72}$$

for some $a > 0$. *Then* ϕ *is not a cyclic vector.*

PROOF. Let D denote the disc $|z - ½| < ½$. Fix p $(0 < p < ½)$. Using (67) it can be shown that the restriction map $f \longrightarrow f|D$ is a bounded linear transformation from the Bergman space into $H^p(D)$. Hence, if ϕ were cyclic in the Bergman space, then $\phi|D$ would be cyclic in $H^p(D)$. But using (72) it can be shown that $\phi|D$ has a singular inner factor (whose associated singular measure has a mass point at $z = 1$). This is a contradiction.

Hedberg's proof of Proposition 41 is somewhat similar to the methods used in his paper **[38]**. His proof also shows that if G is a plane domain of finite area and finite connectivity, then $H^\infty(G)$ is dense in $L^2_a(G)$.

In connection with Questions 25 and 25′, the following questions seem natural for the Dirichlet space. Recall that the "Dirichlet space" is $H^2(\beta)$, $\beta(n) = (n+1)^{1/2}$.

Question 30. If ϕ is in the Dirichlet space, and if

$$|\phi(z)| \geqslant [\log 1/(1 - |z|)]^{-1/2},$$

is ϕ a cyclic vector? In particular, if ϕ and $1/\phi$ are in the Dirichlet space, is ϕ a cyclic vector?

More information on the role of inner functions in spaces like the Bergman space (but with a weighted L^1 area norm) may be found in Duren, Romberg, and Shields **[25**, §6, pp. 54–58**]**.

For cyclic vectors of the backward shift see Douglas, Shapiro, and Shields **[22]**, and Hilden and Wallen **[45]**.

Proposition 42 and Question 27 are due to Herrero **[43]**.

In Appendix I to his thesis **[58]** Lambert shows that if two injective unilateral shifts are quasi-similar, then they are similar. Using shifts with zero weights, T. Hoover shows that a compact quasi-nilpotent shift can be quasi-similar to a shift whose spectrum is the closed unit disc (see **[47**, pp. 16, 17**]**).

For weighted shifts of higher multiplicity (i.e., shifts with operator weights) see Lambert **[61]**, **[62]**, Lambert and Turner **[63]**, and Nikolskiĭ **[70]**.

In conclusion we mention two problems.

Question 31. Find reasonable necessary and sufficient conditions for a power series with nonnegative coefficients to be in $H^\infty(\beta)$.

This seems difficult in the case of the Dirichlet space, for example. See G. D. Taylor **[95]** for partial results. The next question has been around for some time.

Question 32. If a transitive algebra of operators contains an injective unilateral shift, must it be strongly dense in $L(H)$?

Arveson **[3]** showed that this is true for the unweighted unilateral shift. Lambert established this for strictly cyclic shifts **[60]**. See also Rickart **[78]**, where a more general theorem is proved.

Added in proof (April 1974). Question 1 of §4 and Question 8 of §7 have been answered. Charles Horowitz in his dissertation, University of Michigan (1974), shows that the union of two zero sets in the Bergman space need not be a zero set. Hence there are two invariant subspaces in the Bergman space whose intersection is trivial. He also shows that if $\{z_n\}$ is a subset of a zero set, then there is a function in the space that vanishes exactly at these points and nowhere else in the open unit disc. These results also hold for many other subnormal weighted shifts.

The special case of Question 6 (§6) that concerns the Dirichlet space has been solved by L. Carleson. We give his result.

PROPOSITION 44 (CARLESON). *Let* $\beta(n) = (n + 1)^{1/2}$. *Let* $\phi \in H^\infty(\beta)$, *with* $|\phi(z)| \geqslant d > 0$ $(|z| < 1)$. *Then* $1/\phi \in H^\infty(\beta)$.

PROOF (OUTLINE). Let f be a function with a finite Dirichlet integral. Then f, $\phi f \in H^2(\beta)$, and we must show that $f/\phi \in H^2(\beta)$. The hypotheses on ϕ imply that ϕ is an outer function.

Carleson **[16b]** has given a representation for the Dirichlet integral in terms of boundary integrals involving the inner-outer factorization. An examination of his formula shows that the inner factor of f/ϕ does not cause any trouble.

Let $u(t) = \log|f(e^{it})|$, $v(t) = \log|\phi(e^{it})|$. Replacing ϕ by ϕ/d we may assume that $|\phi| \geqslant 1$, and so $v \geqslant 0$. Let

$$G(u; t, s) = (u(t + s) - u(t))(e^{2u(t+s)} - e^{2u(t)}). \tag{73}$$

Note that $G \geqslant 0$, since the two factors in the product have the same sign. An examination of Carleson's formula shows that the proof will be complete if we can prove the existence of a constant c such that

$$G(u - v; t, s) \leqslant c(G(u; t, s) + G(u + v; t, s)), \qquad -\pi \leqslant t, s \leqslant \pi. \tag{74}$$

(We are using the fact that both f and ϕf have a finite Dirichlet integral.)

Fix t, s, so that the boundary values of f and ϕ exist at $\exp(it)$ and at $\exp(i(t+s))$. Let $A(u) = u(t+s) - u(t)$. If we factor out $\exp(2u(t))$ in (73) we have

$$G(u; t, s) = e^{2u(t)}A(u)(e^{2A(u)} - 1). \tag{75}$$

Note that A is linear: $A(u+v) = A(u) + A(v)$. Let $x = A(u)$, $a = A(v)$, and $F(t) = t(e^{2t} - 1)$. Let $|\phi| \leqslant M$. Then one shows that $|a| \leqslant \log M$. Since $v \geqslant 0$ we have $u - v \leqslant u \leqslant u + v$. A little checking now shows that to prove (74) it suffices to prove the existence of a constant c such that

$$F(x-a) \leqslant c(F(x) + F(x+a)), \qquad -\infty < x < \infty; \ |a| \leqslant \log M. \tag{76}$$

We omit the details of this proof. Q.E.D.

The following problem was raised by Hong Wha Kim.

Question 33. If T is a hyponormal unilateral shift and if p is a polynomial, must $p(T)$ be hyponormal?

If correct this would mean that $H^\infty(\beta)$ consists entirely of hyponormal operators. The same question can be asked for hyponormal bilateral shifts and Laurent polynomials.

Finally, we give the proof that H^∞ is the dual (in the isomorphic theory) of two nonisomorphic Banach spaces (see the end of §4, preceding Question 2′). This proof was shown to me by Haskell Rosenthal.

By a result of J. Lindenstrauss [**64**a] there is a Banach space X (a subspace of l') with the following two properties. First, X^* is isomorphic to l^∞. Second, X is not isomorphic to a complemented subspace of a conjugate Banach space.

We also have Per Beurling's result [**16**a] that H^∞ is isomorphic to $H^\infty \oplus l^\infty$. Hence H^∞ is isomorphic to the dual of $L'/H' \oplus X$. We shall show that this space is not isomorphic to L'/H'. Note that if A denotes the disc algebra (the closure of the polynomials in the continuous functions on the circle) then $A^* = L'/H' \oplus M_s$, where M_s denotes the singular measures (with respect to Lebesgue measure) on the circle. Thus if L'/H' were isomorphic to $L'/H' \oplus X$, then X would be isomorphic to a complemented subspace of a conjugate space, which is a contradiction.

Note that we also have the following preduals of H^∞: L'/H', $L'/H' \oplus l'$, $L'/H' \oplus L'$. Are these three spaces all isomorphic to one another?

BIBLIOGRAPHY

1. D. Aharonov, H. S. Shapiro and A. L. Shields, *Weakly invertible elements in the space of square-summable holomorphic functions*, J. London Math. Soc. (to appear).

2. N. Aronszajn, *Theory of reproducing kernels*, Trans. Amer. Math. Soc. **68** (1950), 337–404. MR **14**, 479.

3. William B. Arveson, *A density theorem for operator algebras*, Duke Math. J. **34** (1967), 635–647. MR **36** #4345.

4. V. Bargmann, *On a Hilbert space of analytic functions and an associated integral transform*, Comm. Pure Appl. Math. **14** (1961), 187–214. MR **28** #486.

5. ————, *Remarks on a Hilbert space of analytic functions*, Proc. Nat. Acad. Sci. USA **48** (1962), 199–204. MR **24** #A2845.

6. David Beckles, *Banach algebras with a single generator* (preprint).

7. ————, *An algebra with nested ideals* (preprint).

8. C. A. Berger and J. G. Stampfli, *Mapping theorems for the numerical range*, Amer. J. Math. **89** (1967), 1047–1055. MR **36** #5744.

9. S. Bergman, *The kernel function and conformal mapping*, Math. Surveys, vol. 5, Amer. Math. Soc., Providence, R. I., 1950. MR **12**, 402.

10. Cz. Bessaga and A. Pełczyński, *Spaces of continuous functions.* IV. *On isomorphical classification of spaces of continuous functions*, Studia Math. **19** (1960), 53–62. MR **22** #3971.

11. ————, *On extreme points in separable conjugate spaces*, Israel J. Math. **4** (1966), 262–264. MR **35** #2126.

12. Arne Beurling, *On two problems concerning linear transformations in Hilbert space*, Acta Math. **81** (1949), 239–255.

13. ————, *A critical topology in harmonic analysis on semigroups*, Acta Math. **112** (1964), 215–228. MR **29** #6255.

14. Garrett Birkhoff, *Lattice theory*, rev. ed., Amer. Math. Soc. Colloq. Publ., vol. 35, Amer. Math. Soc., Providence, R. I., 1961. MR **23** #A815.

15. Lennart Carleson, *Sets of uniqueness for functions regular in the unit circle*, Acta Math. **87** (1952), 325–345. MR **14**, 261.

16. ————, *Interpolations by bounded analytic functions and the corona problem*, Ann. of Math. (2) **76** (1962), 547–559. MR **25** #5186.

16a. ————, *Interpolations by bounded analytic functions and the corona problem*, Proc. Internat. Congr. Mathematicians (Stockholm, 1962), 314–316. Inst. Mittag-Leffler, Djursholm (1963). MR **31** #549.

16b. ————, *A representation formula for the Dirichlet integral*, Math. Z. **73** (1960), 190–196. MR **22** #3803.

17. ———, *The corona theorem*, Proc. Fifteenth Scandinavian Congress (Oslo, 1968), Lecture Notes in Math., vol. 118, Springer-Verlag, Berlin, 1970, pp. 121–132. MR **41** #8696.

18. I. W. Cohen and N. Dunford, *Transformations on sequence spaces*, Duke Math. J. **3** (1937), 689–701.

19. James A. Deddens, *Reflexive operators*, Indiana Univ. Math. J. **20** (1971), 887–889.

20. Karel de Leeuw, *Fourier series of operators and an extension of the F. and M. Riesz theorem*, Bull. Amer. Math. Soc. **79** (1973), 342–344.

21. W. F. Donoghue, Jr., *The lattice of invariant subspaces of a completely continuous quasi-nilpotent transformation*, Pacific J. Math. **7** (1957), 1031–1035. MR **19**, 1066.

22. R. G. Douglas, H. S. Shapiro and A. L. Shields, *Cyclic vectors and invariant subspaces for the backward shift operator*, Ann. Inst. Fourier (Grenoble) **20** (1970), fasc. 1, 37–76. MR **42** #5088.

23. N. Dunford, and J. T. Schwartz, *Linear operators*. Part I: *General theory*, Pure and Appl. Math., vol. 7, Interscience, New York, 1958. MR **22** #8302.

24. P. L. Duren, *Theory of H^p spaces*, Pure and Appl. Math., vol. 38, Academic Press, New York, 1970. MR **42** #3552.

25. P. L. Duren, B. W. Romberg and A. L. Shields, *Linear functionals on H^p spaces with $0 < p < 1$*, J. Reine Angew. Math. **238** (1969), 32–60. MR **41** #4217.

26. Ciprian Foiaş and James P. Williams, *Some remarks on the Volterra operator*, Proc. Amer. Math. Soc. **71** (1972), 177–184. MR **45** #4194.

27. Ralph Gellar, *Shift operators in Banach space*, Dissertation, Columbia University, New York, N. Y., 1968.

28. ———, *Operators commuting with a weighted shift*, Proc. Amer. Math. Soc. **23** (1969), 538–545. MR **41** #4277a.

29. ———, *Cyclic vectors and parts of the spectrum of a weighted shift*, Trans. Amer. Soc. **146** (1969), 69–85. MR **41** #4277b.

30. ———, *Two sublattices of weighted shift invariant subspaces* (preprint).

31. Ralph Gellar and Domingo Herrero, *Hyperinvariant subspaces of bilateral weighted shifts*, Indiana Univ. Math. J. (to appear).

32. R. Gellar and L. J. Wallen, *Subnormal weighted shifts and the Halmos-Bram criterion*, Proc. Japan Acad. **46** (1970), 375–378.

33. Sandy Grabiner, *Weighted shifts and Banach algebras of power series*, Amer. J. Math. (to appear).

34. P. R. Halmos, *A Hilbert space problem book*, Van Nostrand, Princeton, N. J., 1967. MR **34** #8178.

35. ———, *Ten problems in Hilbert space*, Bull. Amer. Math. Soc. **76** (1970), 887–933. MR **42** #5066.

36. K. J. Harrison, *On the unicellularity of weighted shifts*, J. Austral. Math. Soc. **12** (1971), 342–350. MR **46** #9780.

37. ———, *A new attainable lattice* (preprint).

38. Lars Inge Hedberg, *Weighted mean approximation in Carathéodory regions*, Math. Scand. **23** (1968), 113–122. MR **41** #2028.

39. Henry Helson, *Lectures on invariant subspaces*, Academic Press, New York, 1964. MR **30** #1409.

40. ———, *Invariant subspaces of the weighted shift*, Colloq. Math. Soc. Janos Bolyai. 5. Hilbert space operators. Tihany, Hungary, 1970.

41. Domingo A. Herrero, *Subnormal bilateral weighted shifts* (preprint).

42. ———, *On the invariant subspace lattice* $1 + \omega^*$, Acta Sci. Math. (Szeged) **35** (1973), 217–223.

43. ———, *Eigenvectors and cyclic vectors for bilateral weighted shifts* (preprint).

44. D. A. Herrero and A. L. Lambert, *On strictly cyclic algebras, P-algebras, and reflexive operators* (preprint).

45. H. M. Hilden and L. J. Wallen, *Some cyclic and non-cyclic vectors of certain operators* (preprint).

46. K. Hoffman, *Banach spaces of analytic functions*, Prentice-Hall Ser. in Modern Analysis, Prentice-Hall, Englewood Cliffs, N. J., 1962. MR **24** #A2844.

47. Thomas Benton Hoover, *Quasi-similarity and hyperinvariant subspaces*, Dissertation, University of Michigan, Ann Arbor, Mich., 1970.

48. Lars Ingelstam, *Hilbert algebras with identity*, Bull. Amer. Math. Soc. **69** (1963), 794–796. MR **27** #4096.

49. M. V. Keldyš, *Sur l'approximation en moyenne par polynômes des fonctions d'une variable complexe*, Mat. Sb. **16 (58)** (1945), 1–20. MR **7**, 64.

50. Robert Lee Kelley, *Weighted shifts on Hilbert space*, Dissertation, University of Michigan, Ann Arbor, Mich., 1966.

51. Edward Kerlin and Alan L. Lambert, *Strictly cyclic shifts on* l_p, Acta Sci. Math. (Szeged) **35** (1973), 87–94.

52. V. L. Klee, Jr., *Some characterizations of reflexivity*, Revista Ci. Lima **52** (1950), nos. 3–4, 15–23. MR **13**, 250.

53. B. I. Korenblium, *Quasianalytic classes of functions in a circle*, Dokl. Akad. Nauk SSSR **164** (1965), 36–39 = Soviet Math. Dokl. **6** (1965), 1155–1158. MR **35** #3074.

54. ———, *Functions holomorphic in a disk and smooth in its closure*, Dokl. Akad. Nauk SSSR **200** (1971), 24–27 = Soviet Math. Dokl. **12** (1971), 1312–1315. MR **44** #4212.

55. ———, *Invariant subspaces of the shift operator in weighted Hilbert space*, Mat. Sb. **89 (131)** (1972), 110–137 = Math. USSR Sb. **18** (1972), 111–138.

56. ———, *Closed ideals in the ring* A^n, Funkcional. Anal. i Priložen. **6** (1972), 38–53. (Russian)

57. R. H. Kulkarni, *Subnormal operators and weighted shifts*, Dissertation, Indiana University, Bloomington, Ind., 1970.

58. Alan Leslie Lambert, *Strictly cyclic operator algebras*, Dissertation, University of Michigan, Ann Arbor, Mich., 1970.

59. ———, *Strictly cyclic weighted shifts*, Proc. Amer. Math. Soc. **29** (1971), 331–336. MR **43** #970.

60. ———, *Strictly cyclic operator algebras*, Pacific J. Math. **39** (1971), 717–726. MR **46** #9762.

61. ———, *The algebra generated by an invertibly weighted shift*, J. London Math. Soc. **5** (1972), 741–747.

62. Alan Leslie Lambert, *Unitary equivalence and reducibility of invertibly weighted shifts*, Bull. Austral. Math. Soc. **5** (1971), 157–175. MR **45** #4196.

63. A. L. Lambert and T. R. Turner, *The double commutants of invertibly weighted shifts*, Duke Math. J. **39** (1972), 385–389. MR **46** #9781.

64. E. Landau, *Darstellung and Begründung einiger neuerer Ergebnisse der Funktionentheorie*, Berlin, 1929.

64a. J. Lindenstrauss, *On a certain subspace of* l', Bull Acad. Polon. Sci. Sér. Math., Astronom., Phys. **12** (1964), 539–542. MR 30 #5153.

65. A. I. Markuševič, *Theory of functions of a complex variable*. Vol. III, GITTL, Moscow, 1950; English transl., Prentice-Hall, Englewood Cliffs, N. J., 1967. MR **12**, 87; **35** #6799.

66. A. A. Miljutin, *Isomorphism of the spaces of continuous functions over compact sets of the cardinality of the continuum*, Teor. Funkciĭ Funkcional. Anal. i. Priložen Vyp. **2** (1966), 150–156. (Russian) MR **34** #6513.

67. James D. Nelson, *A characterization of zero sets for* A^{∞}, Michigan Math. J. **18** (1971), 141–147. MR **44** #410.

68. D. J. Newman and H. S. Shapiro, *Fischer spaces of entire functions*, Proc. Sympos. Pure Math., vol. 11, Amer. Math. Soc., Providence, R. I., 1968, pp.360–369. MR **38** #2333.

69. N. K. Nikol′skiĭ, *The invariant subspaces of certain completely continuous operators*, Vestnik Leningrad. Univ. **20** (1965), no. 7, 68–77. (Russian) MR **32** #2911.

70. ————, *On invariant subspaces of weighted shift operators*, Mat. Sb. **74** **(116)** (1967), 171–190 = Math. USSR Sb. **3** (1967), 159–176. MR **37** #4659.

71. ————, *Basicity and unicellularity of weighted shift operators*, Izv. Akad. Nauk SSSR Ser. Mat. **32** (1968), 1123–1137 = Math. USSR Izv. **2** (1968), 1077–1090. MR **38** #6374.

72. ————, *Spectral synthesis for a shift operator and zeros in certain classes of analytic functions smooth up to the boundary*, Dokl. Akad. Nauk SSSR **190** (1970), 780–783 = Soviet Math. Dokl. **11** (1970), 206–209. MR **41** #5999.

73. ————, *Spectral synthesis and the problem of weighted approximation in some spaces of analytic functions that grow near the boundary*, Izv. Akad. Nauk Armjan. SSR Ser. Mat. **6** (1971), no. 5, 345–367. (Russian) MR **45** #7504.

74. Eric A. Nordgren, *Closed operators commuting with a weighted shift*, Proc. Amer. Math. Soc. **24** (1970), 424–428. MR **41** #2435.

74a. A. Pełczyński, *On the isomorphism of the spaces m and M,* Bull. Acad. Polon. Sci. Sér. Sci. Math. Astr. Phys. **6** (1958), 695–696. MR **21** #1513.

75. G. Pólya and G. Szegö, *Aufgaben und Lehrsätze aus der Analysis*, Vol. I, Springer, Berlin, 1925.

76. Calvin R. Putnam, *On normal operators in Hilbert space*, Amer. J. Math. **73** (1951), 357–362. MR **12**, 717.

77. M. Rabindranathan, *On cyclic vectors of weighted shifts*, Proc. Amer. Math. Soc. (to appear).

77a. H. Radjavi and P. Rosenthal, *Invariant subspaces*, Springer-Verlag, Berlin-Heidelberg-New York, 1973.

78. Charles E. Rickart, *The uniqueness of norm problem in Banach algebras*, Ann. of Math. (2) **51** (1950), 615–628. MR **11**, 670.

79. William C. Ridge, *Approximate point spectrum of a weighted shift*, Trans. Amer. Math. Soc. **147** (1970), 349–356. MR **40** #7843.

80. Peter M. Rosenthal, *On lattices of invariant subspaces*, Dissertation, University of Michigan, Ann Arbor, Mich., 1967.

80a. ———, *Remarks on invariant subspace lattices*, Canad. Math. Bull. **12** (1969), 639–643. MR **40** #3346.

80b. ———, *A note on unicellular operators*, Proc. Amer. Math. Soc. **19** (1968), 505–506. MR **36** #5753.

80c. ———, *Examples of invariant subspace lattices*, Duke Math. J. **37** (1970), 103–112. MR **40** #4797.

81. Emile Boyd Roth, *Conjugate space representations of Banach spaces*, Pacific J. Math **32** (1970), 793–797. MR **41** #2363.

82. Lee A. Rubel and Allen L. Shields, *The space of bounded analytic functions on a region*, Ann. Inst. Fourier (Grenoble) **16** (1966), fasc. 1, 235–277. MR **33** #6440.

83. W. Rudin, *Real and complex analysis*, McGraw-Hill, New York, 1966. MR **35** #1420.

84. Harold S. Shapiro, *Weakly invertible elements in certain function spaces, and generators in* l_1, Michigan Math. J. **11** (1964), 161–165. MR **29** #3620.

85. ———, *Weighted polynomial approximation and boundary behaviour of analytic functions*, Contemporary Problems in Theory Anal. Functions (Internat. Conf., Erevan, 1965), "Nauka", Moscow, 1966, pp. 326–335. (Russian) MR **35** #383.

86. ———, *Some remarks on weighted polynomial approximations of holomorphic functions*, Mat. Sb. **73 (115)** (1967), 320–330 = Math. USSR Sb. **2** (1967), 285–294. MR **36** #395.

87. H. S. Shapiro and A. L. Shields, *On the zeros of functions with finite Dirichlet integral and some related function spaces*, Math. Z. **80** (1962), 217–229. MR **26** #2617.

88. A. L. Shields and L. J. Wallen, *The commutants of certain Hilbert space operators*, Indiana Univ. Math. J. **20** (1970), 777–788. MR **44** #4558.

89. Ivan Singer, *On a theorem of J. D. Weston*, J. London Math. Soc. **34** (1959), 320–324. MR **21** #5884.

90. M. F. Smiley, *Real Hilbert algebras with identity*, Proc. Amer. Math. Soc. **16** (1965), 440–441. MR **31** #591.

91. J. G. Stampfli, *Which weighted shifts are subnormal?*, Pacific J. Math. **17** (1966), 367–379. MR **33** #1740.

92. B. Sz.-Nagy and C. Foiaş, *Analyse harmonique des opérateurs de l'espace de Hilbert*, Masson, Paris; Akad. Kiadó, Budapest, 1967; English rev. transl., North-Holland, Amsterdam; American Elsevier, New York; Akad. Kiadó, Budapest, 1970. MR **37** #778; **43** #947.

93. B. A. Taylor and D. L. Williams, *Ideals in rings of analytic functions with smooth boundary values*, Canad. J. Math. **22** (1970), 1266–1283. MR **42** #7905.

94. B. A. Taylor and D. L. Williams, *Zeros of Lipschitz functions analytic in the unit disc*, Michigan Math. J. **18** (1971), 129–139. MR **44** #409.

95. G. D. Taylor, *Multipliers on* D_α, Trans. Amer. Math. Soc. **123** (1966), 229–240. MR **34** #6514.

96. Thomas Rolf Turner, *Double commutants of singly generated operator algebras*, Dissertation, University of Michigan, Ann Arbor, Mich., 1971.

97. James P. Williams, *Minimal spectral sets of compact operators*, Acta Sci. Math. (Szeged) **28** (1967), 93–106. MR **36** #725.

98. A. Zygmund, *Trigonometric series*. Vol. I, 2nd ed., Cambridge Univ. Press, New York, 1959. MR **21** #6498.

UNIVERSITY OF MICHIGAN

Mathematical Surveys
Volume 13
1974

III

A VERSION OF MULTIPLICITY THEORY

BY

ARLEN BROWN

AMS (*MOS*) *subject classifications* (1970). Primary 47–02, 47B15.

Preparation of this paper was supported in part by a grant from the National Science Foundation.

Introduction. If A and B are (bounded linear) operators on Hilbert spaces H and K, respectively, and if there exists a unitary transformation ϕ of H onto K such that $B = \phi A \phi^*$ (briefly, if A and B are *unitarily equivalent* operators), then there is no possible criterion based on the geometry of Hilbert space alone by means of which A can be distinguished from B; they are abstractly the same operator. It is natural to look for characteristics of operators on Hilbert space that would enable one to determine whether two given operators are unitarily equivalent, for this is just a continuation into the infinite-dimensional situation of the lines of investigation in the matrix theory of the 19th century that culminated in the various familiar canonical forms and systems of invariants. The task of finding a "complete set of unitary invariants" for all operators on Hilbert space, however, is lengthy and arduous. Indeed, it is a fairly difficult business to give such a set of invariants for $n \times n$ matrices (see, for example, **[8]**, **[12]**, **[9]**, and **[10]**). Moreover, the methods used thus far in the finite-dimensional case do not offer the slightest hope of generalizing to the infinite-dimensional case except for a very modest number of quite special classes of operators (see, for example, **[2]**, **[3]**, and **[10]**).

On the other hand, for selfadjoint, or normal, matrices the story is very much simpler; a complete set of unitary invariants for a normal matrix is provided by its spectrum "counting multiplicities." It was natural to hope that a suitable generalization of this result would be valid for normal operators on Hilbert space, and this is, indeed, the case, as was first shown by Hellinger in 1907 **[6]**. (In actual fact, Hellinger limited his attention to the selfadjoint case, but that is a minor technicality.) The problem of how to "count" spectral "multiplicities" for a normal operator on an infinite-dimensional Hilbert space can be viewed in various different contexts, and has therefore been solved in various different ways. One point point of view goes as follows: A normal operator is, by definition, one that can inhabit an abelian von Neumann algebra. If one looks at normal operators in this light, then one will naturally view multiplicity theory as a portion of the structure theory of abelian von Neumann algebras and, beyond that, of type I algebras in general. For an account of this view of things the reader may consult the original treatment of Segal [11] or Dixmier's treatise [4]. Another point of attack is to regard multiplicity theory as a part of the structure theory of C^*-algebras; for treatments along these lines see [1] or [7]. Still another point of view, and the one followed in this note, is to try to stick as close as possible to the elementary methods that work in the finite-dimensional case. This turns out to be entirely possible, though it is essential to replace the finite subsets of $\mathbf{C}$, which suffice to describe the most general normal finite matrix, by something more general–usually measures on $\mathbf{C}$, as in Hellinger's original memoir [6]. Moreover (another technical point), Hellinger's way of counting multiplicities explicitly assumed the underlying Hilbert space to be separable, so that some improvement over his solution of the problem, at least in technique, had to be found if the problem was to be solved in the general, nonseparable case.

In was Wecken [13], in 1939, who first pointed out that, in order to describe the general normal operator (or spectral measure, or abelian von Neumann algebra) on a nonseparable Hilbert space, one needs to replace measures by σ-*ideals* of measures. The present treatment follows Wecken's development in all essentials. Another version of multiplicity theory that has greatly influenced this treatment, however, is that of Halmos [5], from which we have borrowed not only some notation and terminology, but also a point of view. An unusual feature of the treatment presented here is the introduction of the Hilbert spaces $L_2(J)$. These spaces are also employed by Edward Nelson in [9].

In what follows all Hilbert spaces are complex, all subspaces are closed, and all operators are linear and bounded (although this last restriction is wholly unnecessary, and is imposed solely in order to avoid distractions). If μ and ν are measures on a measurable space $(\mathbf{M}, S)$, then we write $\mu \prec \nu$ to indicate that μ is absolutely continuous with respect to ν, and $\mu \sim \nu$ to indicate that μ and ν are equivalent, i.e., that $\mu \prec \nu$ and $\nu \prec \mu$.

If $(\mathbf{M}, S)$ is a measurable space and H is a Hilbert space, then by a *spectral measure* from $(\mathbf{M}, S)$ to $L(H)$ we shall mean, as is customary, a set function E assigning to each measurable set $M \in S$ a projection $E(M)$ on H in such a way that (i) $E(\mathbf{M}) = 1$, and (ii) E is countably additive in the sense that $E(M)x = \Sigma_n E(M_n)x$ for every $x \in H$ whenever $\{M_n\}$ is a countable, disjoint sequence of measurable sets with union M. (In the sequel the underlying measurable space $(\mathbf{M}, S)$ will play almost no role, and we shall ordinarily refer to E simply, if somewhat illogically, as a spectral measure *on* H.) If x, y is a pair of vectors in H, then the complex measure $(E(M)x, y)$ will be denoted by $\mu_{x,y}$, except that, when $x = y$, we shall write μ_x instead of $\mu_{x,x}$. We suppose the reader familiar with the elementary properties of spectral measures.

Specifically, we shall make free use of the spectral integrals $\int f dE = \int_{\mathbf{M}} f dE$ of those measurable, complex valued functions f that are essentially bounded $[E]$. Likewise, we assume the fact that the representation $f \longrightarrow \int f dE$ is an isometric *-isomorphism of the Banach *-algebra $L_\infty[E]$ onto a norm closed, commutative *-subalgebra of $\mathcal{L}(\mathcal{H})$, and that all of the operators $\int f dE$ *doubly commute* E, i.e., commute with every operator that commutes with (the range of) E.

Like any integral, the spectral integral $\int f dE$ may be defined or characterized in various slightly different ways. For our purposes the simplest and quickest approach will suffice: The bilinear functional of $\int f dE$ is given by the equation $(\int f dEx, y) = \int f d\mu_{x,y}$. It is, of course, possible to define $\int f dE$ as a (closed, normal, densely defined) unbounded operator even for unbounded measurable functions f, but we shall have no occasion to make use of this refinement.

In what follows it is the spectral measure that consistently occupies center stage. In particular, it is for spectral measures that we obtain unitary invariants (Theorem 8.5). But it should be remembered that these invariants (and, indeed, all of the structure theory accompanying them) translate at once into the language of a single normal operator. This is because of the *spectral theorem*, by which we here mean the following fact.

Let A *be a normal operator on a Hilbert space* $\mathcal{H}$. *Then there exists a unique spectral measure* E_A *from the complex plane* $\mathbf{C}$ *(equipped with its ring of Borel sets) to* $\mathcal{L}(\mathcal{H})$ *such that* $\int_{\mathbf{C}} \lambda dE_A = A$. *This spectral measure, to be called the* spectral measure of A, *has the further properties that it is supported by the spectrum* Λ *of* A *(i.e., that* $E_A(\mathbf{C}\backslash\Lambda) = 0$*), that the spectral integral* $\int f dE_A$ *coincides with* $f(A)$ *for every bounded Baire function* f *on* Λ, *and that an operator* $T \in \mathcal{L}(\mathcal{H})$ *commutes with* A *(and* A^**) if and only if it commutes with (the range of)* E.

If E and F are two spectral measures on the same carrier $(\mathbf{M}, \mathcal{S})$ taking values in $\mathcal{L}(\mathcal{H})$ and $\mathcal{L}(\mathcal{K})$, respectively, then E and F are *unitarily equivalent* if there exists a unitary mapping Φ of $\mathcal{H}$ onto $\mathcal{K}$ such that $\Phi^* F(M)\Phi = E(M)$ for every $M \in \mathcal{S}$. Because of the uniqueness of the spectral measure E_A in the spectral theorem, it is clear that two normal operators are unitarily equivalent if and only if their spectral measures are.

1. The standard spectral measure. Let $(\mathbf{M}, \mathcal{S}, \mu)$ be a finite measure space, and let $\mathcal{H} = L_2[\mu]$. For each f in $L_\infty[\mu]$ the mapping $g \longrightarrow fg$ of $\mathcal{H}$ into itself is an operator on $\mathcal{H}$ which we shall denote by A_f. If we make the assignment $E(M) = A_{\chi_M}$ for every $M \in \mathcal{S}$, then E is obviously a spectral measure.

We shall refer to such a spectral measure as a *standard* one. A simple check shows that if $f \in L_\infty[\mu]$, then $\int f dE = A_f$. We observe that if, in particular, μ is a Borel measure on $\mathbf{C}$, then E is the spectral measure of multiplication by λ–the so-called *position operator*. We have the following uniqueness theorem.

THEOREM 1.1. *Let μ and ν be two finite measures on $(\mathbf{M}, \mathcal{S})$. Then the standard spectral measures on $L_2[\mu]$ and $L_2[\nu]$ are unitarily equivalent if and only if $\mu \sim \nu$. Moreover, such a unitary mapping is necessarily implemented by a multiplication.*

PROOF. Suppose that Φ is a unitary mapping of $L_2[\mu]$ onto $L_2[\nu]$ such that

$$A_{\chi_M} \text{ (on } L_2[\mu]) = \Phi^* A_{\chi_M} \text{ (on } L_2[\nu])\Phi$$

for every measurable set M. Then, as is easily seen,

$$A_f \text{ (on } L_2[\mu]) = \Phi^* A_f \text{ (on } L_2[\nu])\Phi$$

for every bounded measurable function f on $\mathbf{M}$. It follows at once that, if ϕ denotes the function $\Phi(1)$, then Φ agrees with multiplication by ϕ on the bounded functions in $L_2[\mu]$. But then Φ is multiplication by ϕ. (This is a standard argument: Multiplication by ϕ is a closed transformation and agrees with the bounded operator Φ on a dense set.) Since Φ is an isometry, we have

$$\|\Phi(\chi_M)\|^2 = \int_M |\phi|^2 \, d\nu = \|\chi_M\|^2 = \mu(M)$$

for every measurable set M, which shows that $\mu \prec \nu$. (ϕ belongs to $L_2[\nu]$ because $\int |\phi|^2 d\nu = \mu(\mathbf{M}) < +\infty$.) But then, by symmetry, we also have $\nu \prec \mu$, and therefore $\mu \sim \nu$.

Suppose on the other hand that $\mu \sim \nu$ and that μ is the indefinite integral $[\nu]$ of the function ψ. Set $\phi = \psi^{1/2}$ and define $\Phi(f) = \phi f$ for every $f \in L_2[\mu]$. Then $\|\Phi(f)\|^2$ (in $L_2[\nu]$) $= \int \psi |f|^2 \, d\nu = \int |f|^2 \, d\mu = \|f\|^2$ (in $L_2[\mu]$), so Φ is an isometry of $L_2[\mu]$ into $L_2[\nu]$. Since $\mu \sim \nu$, the function $\phi \neq 0$ a.e. $[\nu]$. But if $g \in L_2[\nu]$ is orthogonal to the range of Φ, then, in particular,

$$(g, \Phi(\chi_M)) = \int_M g\phi \, d\nu = 0$$

for every measurable set M, so that $g\phi = 0$ a.e. $[\nu]$. Thus $g = 0$ a.e. $[\nu]$. Hence the range of Φ is dense, and is therefore all of $L_2[\nu]$. Finally, it is clear that $A_f = \Phi^* A_f \Phi$ for every bounded function f, and in particular for $f = \chi_M$. Q.E.D.

2. Cyclic subspaces. If E is a spectral measure on $\mathcal{H}$ and if x is a vector, then the smallest subspace of $\mathcal{H}$ that contains x and reduces (the range of) E is called the *cyclic subspace*, or, briefly, the *cycle*, generated by x (with respect to E). We shall denote this subspace by $\mathcal{Z}_x$, and the projection of $\mathcal{H}$ onto $\mathcal{Z}_x$ by Z_x. It is obvious that Z_x may be described as the smallest projection P that commutes with E and satisfies the condition $Px = x$, and that $\mathcal{Z}_x$ is the subspace spanned by the set of vectors $\{E(M)x: M \in S\}$. In the event that $\mathcal{Z}_{x_0} = \mathcal{H}$ for some vector x_0, we say that x_0 is a *cyclic vector* for E. When $E = E_A$ is the spectral measure of some normal operator A, the cyclic subspace $\mathcal{Z}_x$ also coincides with the smallest reducing subspace for A that contains the vector x (for a projection P commutes with E if and only if it commutes with A). Thus in this case $\mathcal{Z}_x$ may equally be described as the closure of the set of all vectors

$$\{p(A, A^*)x: \ p(\lambda, \bar{\lambda}) \ \text{a polynomial}\}.$$

The important role played by the cyclic subspaces in the structure theory of spectral measures stems from the following basic theorem.

THEOREM 2.1. *Let E be a spectral measure on a Hilbert space $\mathcal{H}$, let x be a vector in $\mathcal{H}$, and let $\mathcal{Z}_x$ denote the cycle generated by x. Then there exists a unique unitary mapping Φ of $L_2[\mu_x]$ onto $\mathcal{Z}_x$ implementing a unitary equivalence between the restriction of E to $\mathcal{Z}_x$ and the standard spectral measure on $L_2[\mu_x]$ and satisfying the condition $\Phi(1) = x$.*

PROOF. For every measurable set M we have

$$\|E(M)x\|^2 = \mu_x(M) = \int \chi_M \, d\mu_x.$$

Thus the mapping $\chi_M \longrightarrow E(M)x$ is an isometric mapping of the set of all characteristic functions of measurable sets onto a spanning set of vectors in $\mathcal{Z}_x$ that obviously carries 1 onto x. Moreover, if M and N are two measurable sets, then

$$(E(M)x, E(N)x) = (E(M \cap N)x, x) = \int_{M \cap N} d\mu_x = \int \chi_M \chi_N \, d\mu_x.$$

Hence, by a standard piece of Hilbert space geometry, the isometry $\chi_M \longrightarrow E(M)x$ extends to a unitary mapping Φ of $L_2[\mu_x]$ onto $\mathcal{Z}_x$. Moreover, we have

$$\Phi^* E(M) \Phi(\chi_N) = \Phi^* E(M \cap N)x = \chi_{M \cap N} = \chi_M \chi_N$$

for every pair of measurable sets M and N. Hence $\Phi^* E(M)\Phi$ agrees with multiplication by χ_M on a spanning set in $L_2[\mu_x]$, from which it follows, of course, that $\Phi^*E(M)\Phi$ = multiplication by χ_M, and this for every $M \in S$. Thus existence is established. As for uniqueness, it is clear that two such mappings Φ and Ψ give rise to a unitary operator $\Phi^* \circ \Psi$ on $L_2[\mu_x]$ that commutes with the standard spectral measure and leaves the function 1 fixed. By Theorem 1.1 this implies $\Phi^* \circ \Psi = 1$, so that $\Phi = \Psi$. Q.E.D.

Using this theorem it is a simple matter to dissect any spectral measure into standard pieces.

LEMMA 2.2. *There exist vectors* $\{x_\alpha\}$ *in* H *such that* $H = \Sigma_\alpha \oplus Z_{x_\alpha}$.

PROOF. Zorn's lemma. Q.E.D.

THEOREM 2.3. *For any spectral measure* E *on* H *there exist (finite) measures* $\{\mu_\alpha\}$ *on* $(\mathbf{M}, S)$ *such that* E *is unitarily equivalent to the direct sum of the standard spectral measures on* $\Sigma_\alpha \oplus L_2[\mu_\alpha]$.

PROOF. Clear. Q.E.D.

COROLLARY 2.4. *For any normal operator* A *on* H *there exist (finite) Borel measures* $\{\mu_\alpha\}$ *on* $\mathbf{C}$ *(all supported on the spectrum of* A*) such that* A *is unitarily equivalent to the position operator on* $\Sigma_\alpha \oplus L_2[\mu_\alpha]$.

3. Multiplicity: The problem. It is instructive to analyze the meaning of Corollary 2.4 for a normal $n \times n$ matrix, where everything is simple enough to be seen in detail. Suppose A is a normal $n \times n$ matrix, let $\Lambda = \{\lambda_1, \cdots, \lambda_p\}$ be its spectrum, let E_i denote the eigenspace of λ_i, and let E_i be the projection of H onto E_i, $i = 1, \cdots, p$. Then the spectral measure of A is given by $E_A(M) = \Sigma_{\lambda_i \in M} E_i$ for every Borel set M in $\mathbf{C}$. Consequently, if x is an arbitrary vector, the nonzero vectors in the list $E_1x, E_2x, \cdots, E_px$ form an orthogonal basis for Z_x. Thus the most general cycle for E_A is obtained by selecting one vector out of each of some subset of the spaces E_i, and forming the subspace they span—and it is easily seen from this vantage point what the most general decomposition of A (or of E_A) of the type described in Theorem 2.3 is. Even a rudimentary sense of orderliness suggests looking for a more systematic way of dissecting H. Various such schemes may be invented, but there is one, which we shall now examine, that is an excellent guide to things to come. To begin with, let us write Λ_i for the subset of Λ consisting of those eigenvalues whose multiplicity is i:

$$\Lambda_i = \{\lambda_j \in \Lambda: \dim E_j = i\}.$$

Then the function $m(i) = \Lambda_i$ may be called the *multiplicity function* of A (and of E_A), and a dissection of H into orthogonal cycles with respect to A may be said to "fit" the multiplicity function if it consists of exactly one cycle coinciding with $\sum_{\lambda_i \in \Lambda_1} \oplus E_i$, then of exactly two equivalent cycles whose direct sum coincides with $\sum_{\lambda_i \in \Lambda_2} \oplus E_i$, etc. (This manner of splitting up the spectrum corresponds to ordering matrices from the normal matrix store by asking for an A with such-and-such simple eigenvalues, such-and-such double eigenvalues, etc. It may be envisaged as shown in the following figure.

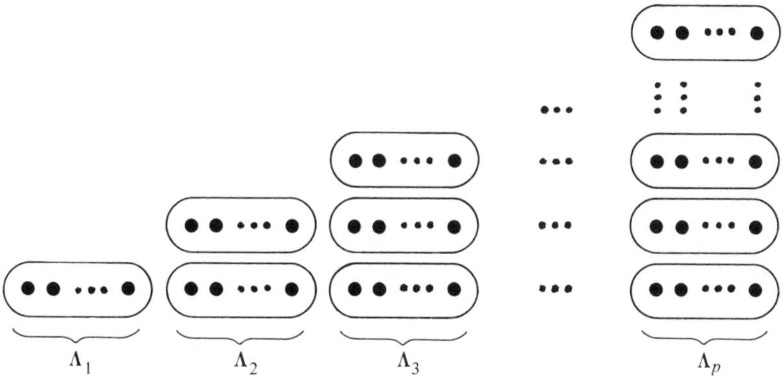

FIGURE 1

Here each dot represents schematically an element of an orthonormal basis for the appropriate eigenspace E_i, and the loops represent suitable cycles.) The following discussion has for its goal the introduction of an analogous orderliness into the dissection into cycles of infinite-dimensional Hilbert spaces with respect to arbitrary spectral measures.

Before turning to the (not altogether entrancing) details of the development, it is worthwhile to pause to consider one other example simple enough so that the cyclic subspaces are all readily accessible. Let μ be a finite measure on $(\mathbf{M}, S)$, and let E denote the standard spectral measure on $L_2[\mu]$. Then for any f in $L_2[\mu]$ it is easily verified that g belongs to Z_f if and only if g vanishes a.e. that f does, which occurs if and only if $\mu_g \prec \mu_f$. Moreover, it is easy to see that $Z_f \perp Z_{f'}$ if and only if μ_f and $\mu_{f'}$ are supported by disjoint sets, and are therefore mutually singular. Thus, in the case of a standard spectral measure, a decomposition of the type described in Theorem 2.3 corresponds to the choice of a maximal pairwise singular family of measures $\mu_f \prec \mu$.

4. The algebra of finite measures. The preceding example suggests—at least it was intended to suggest—that we are going to need a fairly extensive acquaintance with the family, call it $\mathfrak{M}$, of finite measures on $(\mathbf{M}, S)$. The first thing we note is that $\prec$ is a reflexive and transitive relation on $\mathfrak{M}$ that defines a partial ordering of the quotient set of *equivalence classes* of measures modulo the equivalence relation $\mu \sim \nu$. In this note we adopt the convenient legal fiction that the elements of this partially ordered set are actually measures rather than classes of measures, with the understanding that the various concepts introduced as we go along are defined only up to equivalent measures. We employ the notation $\mu \perp \nu$ to indicate that μ and ν are mutually singular measures. The first fact that we need to know is simple enough.

PROPOSITION 4.1. *If* μ *is any measure in* $\mathfrak{M}$, *then the principal ideal* $(\mu) = \{\nu \in \mathfrak{M}\colon \nu \prec \mu\}$ *is order isomorphic with the measure ring of* μ*–i.e., with the space* S/μ *of equivalence classes of measurable sets, where two sets are regarded as equivalent if they are almost equal* $[\mu]$.

PROOF. If $\nu \prec \mu$, then ν is the indefinite integral of some function ϕ with respect to μ, and if $M_0 = \{\lambda \in \mathbf{M}\colon \phi(\lambda) > 0\}$, then $\nu \sim \mu|M_0$. It is clear that $\mu|M_0 \sim \mu|M_1$ if and only if M_0 and M_1 are almost equal $[\mu]$, and hence that $\nu \leftrightarrow M_0$ establishes a one-to-one correspondence between (μ) and the measure ring of μ. Since this correspondence is order preserving, the proposition is proved. Q.E.D.

The second main fact is even simpler.

PROPOSITION 4.2. *If* $\{\nu_n\}$ *is an arbitrary countable family of measures in* $\mathfrak{M}$, *then there exists a measure* $\mu \in \mathfrak{M}$ *such that all of the measures* ν_n *are contained in* (μ).

PROOF. It clearly suffices to treat the case in which $\nu_n(\mathbf{M}) \succ 0$ for all n, and in this case the measure

$$\mu = \sum_n \frac{1}{2^n} \frac{\nu_n}{\nu_n(\mathbf{M})}$$

does all that is required. Q.E.D.

Suppose now that μ and μ' are two measures in $\mathfrak{M}$, and that ν is any measure such that $\mu, \mu' \prec \nu$. Then we can form $\mu \vee \mu'$ and $\mu \wedge \mu'$ in (ν), and while it would appear that these constructs might depend on the choice of ν, in fact they do not. In the first place, if λ is any measure such that $\lambda \prec \mu, \mu'$, then $\lambda \in (\nu)$ and therefore $\lambda \prec \mu \wedge \mu'$, which shows that $\mu \wedge \mu'$ is uniquely determined (up to equivalence only!) by μ and μ'. But also if $\mu, \mu' \prec \lambda$, then $\mu, \mu' \prec \lambda \wedge \nu$ and therefore

$$\mu \vee \mu' \prec \lambda \wedge \nu \prec \lambda,$$

so that $\mu \vee \mu'$ is also independent of ν. In other words, $\mathfrak{M}$ is a lattice. (Note that in this lattice $\mu \wedge \nu = 0$ if and only if $\mu \perp \nu$.) Much more is true, however. Suppose C is any bounded, nonempty subset of $\mathfrak{M}$, and let ν be a measure such that $C \subset (\nu)$. Then C has a supremum $\bigvee C$ in (ν) (this is because of the "method of exhaustion," dear to every measure theorist's heart) and this too is independent of ν. Indeed, if $C \subset (\lambda)$ for some other measure λ then $C \subset (\lambda \wedge \nu)$ and therefore $\bigvee C \prec \lambda \wedge \nu \prec \lambda$, so that $\bigvee C$ is truly the supremum of C in $\mathfrak{M}$. Similarly, if C is *any* nonempty subset of $\mathfrak{M}$, then C possesses an infimum $\bigwedge C$ in $\mathfrak{M}$ (viz., the supremum of the set of all lower bounds of C).

Note. We recall for use later that if C is a bounded set of measures, then there exists a countable subset $\{\mu_n\}$ of C such that $\bigvee C = \bigvee_n \mu_n$. That is just the way the method of exhaustion works.

We observe next that the lattice $\mathfrak{M}$ is *distributive* (for this may be checked "locally"–i.e., in one principal ideal at a time). Finally, if μ and ν are arbitrary elements of $\mathfrak{M}$, then it is possible to find measures μ' and μ'' such that

$$\mu = \mu' \vee \mu'' \quad \text{while} \quad \mu' \prec \nu, \mu'' \perp \nu, \tag{1}$$

for this may also be verified locally. In this decomposition (the "Lebesgue decomposition" of μ with respect to ν) it is clear that μ' is just $\mu \wedge \nu$, and that μ'' is also uniquely determined up to equivalence. We have proved the following result.

THEOREM 4.3. *The system* $\mathfrak{M}$ *of finite measures on* $(\mathbf{M}, S)$ *is a boundedly complete, σ-complete, distributive, relatively complemented lattice in which* $\mu \perp \nu$ *if and only if* $\mu \wedge \nu = 0$.

Note. Such a system is usually known as a *Boolean algebra*. It is not at all difficult to establish the connection between such a lattice and the more modern notion of a *Boolean ring*, but it would be entirely beside the point to go into that here. Incidentally, the simple sum $\mu + \nu$ serves as a representative of the supremum $\mu \vee \nu$–and a suitably weighted sum such as $\Sigma_n(1/2^n)(\mu_n/\mu_n(\mathbf{M}))$ for a countable supremum $\bigvee_n \mu_n$ of (nonzero) measures. There are no comparably simple representations for infima and relative complements.

5. σ-ideals of measures. Along with individual measures, or, better, equivalence classes of measures, we shall also be interested in various assemblages of measures.

DEFINITION. A subset J of $\mathfrak{M}$ is a *σ-ideal* if (i) $\nu \prec \mu$ and $\mu \in J$ imply $\nu \in J$, and (ii) whenever a countable family $\{\mu_n\}$ of measures belongs to J, the supremum $\bigvee_n \mu_n$ also belongs to J.

Examples of σ-ideals are the lattice $\mathfrak{M}$ itself, the zero ideal (0), and all of the principal ideals (μ). Another simple example can be constructed as follows: Fix a measurable set M and form the collection $\{\mu \in \mathfrak{M}: \mu(\mathbf{M}\backslash M) = 0\}$. We shall denote this σ-ideal by (M). (If M is countable, (M) is principal; otherwise, it need not be and, in fact, *cannot* be principal if S is rich enough to contain all singletons.) The following technical fact will be of importance to us in developing the needed theory of σ-ideals.

LEMMA 5.1. *Every σ-ideal is boundedly complete–i.e., if $\mathcal{J}$ is a σ-ideal in $\mathfrak{M}$ and if $\mathcal{C}$ is any bounded (nonempty) subset of $\mathcal{J}$, then $\bigvee \mathcal{C} \in \mathcal{J}$.*

PROOF. As has already been pointed out, $\mathcal{C}$ contains a countable family $\{\mu_n\}$ such that $\bigvee \mathcal{C} = \bigvee_n \mu_n$. Q.E.D.

From this lemma we immediately obtain the following result.

PROPOSITION 5.2. *If a σ-ideal is contained in a principal ideal, i.e., if it is bounded, then it is itself principal.*

PROOF. If $\mathcal{J}$ is bounded then $\mu_0 = \bigvee \mathcal{J}$ exists, and it is clear that $\mathcal{J} \subset (\mu_0)$. On the other hand, by the lemma, we also have $\mu_0 \in \mathcal{J}$ and consequently $(\mu_0) = \mathcal{J}$. Q.E.D.

Now it is obvious that every intersection of σ-ideals is again a σ-ideal, and hence that every nonempty collection of σ-ideals has both a supremum and an infimum in the system $\hat{\mathfrak{M}}$ of all σ-ideals. In other words, $\hat{\mathfrak{M}}$ is a *complete lattice*. In this lattice, of course, the infimum $\bigwedge_\alpha \mathcal{J}_\alpha$ of a family $\{\mathcal{J}_\alpha\}$ of σ-ideals is just the intersection $\bigcap_\alpha \mathcal{J}_\alpha$. Concerning the formation of the supremum $\bigvee_\alpha \mathcal{J}_\alpha$ we have the following result.

LEMMA 5.3. *Let $\{\mathcal{J}_\alpha\}$ be a nonempty family of σ-ideals in $\mathfrak{M}$. Then $\bigvee_\alpha \mathcal{J}_\alpha$ consists of the set of all the measures of the form $\mu = \bigvee_\alpha \mu_\alpha$ where $\{\mu_\alpha\}$ is a (bounded) collection of measures such that $\mu_\alpha \in \mathcal{J}_\alpha$ for each index α.*

PROOF. It is clear by Lemma 5.1 that every such supremum must belong to $\bigvee_\alpha \mathcal{J}_\alpha$. Hence it suffices to show that this set of measures forms a σ-ideal. Suppose first that $\mu = \bigvee_\alpha \mu_\alpha$, where $\mu_\alpha \in \mathcal{J}_\alpha$ for all α, and that $\nu \prec \mu$. Then $\nu = \bigvee_\alpha (\nu \wedge \mu_\alpha)$, and $\nu \wedge \mu_\alpha \prec \mu_\alpha$ belongs to $\mathcal{J}_\alpha$ for all α. Likewise, if $\mu_n = \bigvee_\alpha \mu_\alpha^{(n)}$, $n = 1, 2, \cdots$, where $\mu_\alpha^{(n)} \in \mathcal{J}_\alpha$ for all α and all n, then

$$\bigvee_n \mu_n = \bigvee_n \left(\bigvee_\alpha \mu_\alpha^{(n)}\right) = \bigvee_\alpha \left(\bigvee_n \mu_\alpha^{(n)}\right),$$

and each measure $\bigvee_\alpha \mu_\alpha^{(n)}$ belongs to $\mathcal{J}_\alpha$ because $\mathcal{J}_\alpha$ is a σ-ideal. Q.E.D.

Let us say of a measure μ in $\mathfrak{M}$ that it is *singular* to a collection C of measures (*notation*: $\mu \perp C$) if $\mu \perp \nu$ for every $\nu \in C$. Note that if $\{\mu_n\}$ is a sequence of measures, each of which is singular to C, and if $\nu \in C$, then $\nu \wedge \bigvee_n \mu_n = \bigvee_n \nu \wedge \mu_n = 0$ so that $\bigvee_n \mu_n \perp C$, whence it is clear that the set $C^\perp = \{\mu \in \mathfrak{M}: \mu \perp C\}$ is always a σ-ideal.

PROPOSITION 5.4 (LEBESGUE DECOMPOSITION WITH RESPECT TO A σ-IDEAL). *If J is a σ-ideal and μ is a measure, then μ can be expressed as $\mu = \mu' \vee \mu''$, where μ', μ'' are measures (unique up to equivalence) such that $\mu' \in J$ while $\mu'' \perp J$.*

PROOF. The ideal $(\mu) \cap J$ is principal: Let $(\mu) \cap J = (\mu')$, and write $\mu = \mu' \vee \mu''$, where $\mu'' \perp \mu'$. Then for $\nu \in J$ we have

$$\mu'' \wedge \nu = \mu'' \wedge (\mu \vee \nu) \prec \mu'' \wedge \mu' = 0,$$

so that $\mu'' \perp J$. Q.E.D.

All of the results of this section may be summarized in a single, neat package.

THEOREM 5.5. *The system $\hat{\mathfrak{M}}$ of σ-ideals in $\mathfrak{M}$ is (in the inclusion ordering) a complete, complemented, distributive lattice, the complement of an ideal J being the ideal $J^\perp$ of all measures singular to J. The mapping $\mu \longrightarrow (\mu)$ maps $\mathfrak{M}$ (in the ordering $\prec$) isomorphically into $\hat{\mathfrak{M}}$.*

PROOF. It has already been observed that $\hat{\mathfrak{M}}$ is a complete lattice and it is clear that the zero and unit elements of $\hat{\mathfrak{M}}$ are (0) and $\mathfrak{M}$, respectively. It is also clear that $J \cap J^\perp = (0)$ for every $J \in \hat{\mathfrak{M}}$ and it is an immediate consequence of Proposition 5.4 that $\mathfrak{M} = J \vee J^\perp$. Hence all that is required to complete the proof is to check the distributive laws and verify the final assertion of the theorem. For the former of these, let $\{J_\alpha\}$ and J be σ-ideals. If a measure μ belongs to $\bigvee_\alpha J_\alpha$, then $\mu = \bigvee_\alpha \mu_\alpha$ where $\mu_\alpha \in J_\alpha$ for all α. If μ also belongs to J, then each μ_α belongs to J too, so $\mu_\alpha \in J \cap J_\alpha$ for all α, and therefore $\mu \in \bigvee_\alpha (J \cap J_\alpha)$. This shows that

$$J \wedge \bigvee_\alpha J_\alpha \subset \bigvee_\alpha (J \wedge J_\alpha)$$

and since the reverse inclusion is clear, this establishes one of the desired distributive laws. To verify the other suppose μ is a measure belonging to all of the ideals $J \vee J_\alpha$, and write $\mu = \mu' \vee \mu''$, where $\mu' \in J, \mu'' \in J^\perp$. Then whenever we express μ in the form $\mu = \mu_0 \vee \mu_\alpha$ with $\mu_0 \in J$ and $\mu_\alpha \in J_\alpha$, we have

$$\mu'' = \mu'' \wedge (\mu_0 \vee \mu_\alpha) = \mu'' \wedge \mu_\alpha \prec \mu_\alpha.$$

Thus $\mu'' \in \bigcap_\alpha J_\alpha$, which shows that

$$\bigcap_\alpha (J \vee J_\alpha) \subset J \vee \bigcap_\alpha J_\alpha,$$

and, once again, the reverse inclusion is trivially valid.

Finally, we note at once that $\mu \longrightarrow (\mu)$ is certainly a one-to-one mapping so that it remains only to check that if $\mathcal{C}$ is any nonempty, bounded collection of measures, then

$$\bigcap \{(\mu) \colon \mu \in \mathcal{C}\} = (\bigwedge \mathcal{C}) \quad \text{and} \quad \bigvee \{(\mu) \colon \mu \in \mathcal{C}\} = (\bigvee \mathcal{C}).$$

The former of these is trivial, while the latter is an immediate consequence of Lemma 5.1. Q.E.D.

6. Cycles and measures. In this section we commence the linkup between the lattice structure of $\mathfrak{M}$ (§§4–5) and the special types of cyclic dissections hinted at in §3. To start with, if μ is a measure belonging to $\mathfrak{M}$, then the measures $\nu \prec \mu$ are precisely those of the form

$$\nu(M) = \int_M \phi \, d\mu, \qquad M \in \mathcal{S},$$

where $\phi \in L_1[\mu]$. (This is just the Radon-Nikodym theorem, which we have already used repeatedly.) But if $f = \phi^{1/2}$, then $f \in L_2[\mu]$ and $\nu(M) = \int_M |f|^2 \, d\mu = \mu_f(M)$ for every measurable set M, where μ_f denotes the measure obtained in the usual way from the standard spectral measure on $L_2[\mu]$. Since it is clear conversely that $\mu_f \prec \mu$, we have proved the following fact.

LEMMA 6.1. *The principal ideal* (μ) *coincides with the set of measures* μ_f, $f \in L_2[\mu]$, *obtained from the standard spectral measure on* $L_2[\mu]$.

Because of Theorem 2.1, this implies at once

PROPOSITION 6.2. *If* E *is a spectral measure on* $\mathcal{H}$ *and* $x \in \mathcal{H}$, *then the set of measures* μ_y, $y \in \mathcal{Z}_x$, *coincides with the principal ideal* (μ_x).

The following fact requires no proof, but will be used several times, so we state it here for convenience of reference.

LEMMA 6.3. *The following conditions are equivalent for any vector* x *and any measurable set* M:

$$E(M)x = 0, \qquad \mu_x(M) = 0, \qquad E(\mathbf{M} \backslash M)x = x.$$

As consequences of this observation we obtain at once the following two useful results.

PROPOSITION 6.4. *If T is any operator on H that commutes with E, then $\mu_{Tx} \prec \mu_x$.*

PROOF. If $\mu_x(M) = 0$, then $E(M)x = 0$. If $T \leftrightarrow E$, then $E(M)Tx = TE(M)x = 0$ too. Q.E.D.

PROPOSITION 6.5. *If $\mu_x \perp \mu_y$, then $\mathsf{Z}_x \perp \mathsf{Z}_y$.*

PROOF. Suppose M is a measurable set such that $\mu_x(M) = 0 = \mu_y(\mathbf{M}\backslash M)$. It follows that $E(M)x = 0$, while $E(M)y = y$, and therefore that x and y are orthogonal. But then, if u and v denote vectors belonging to Z_x and Z_y, respectively, then $\mu_u \prec \mu_x$ and $\mu_v \prec \mu_y$ by Proposition 6.2, and hence $\mu_u \perp \mu_v$, and it follows that u and v are orthogonal too. Q.E.D.

Note. Proposition 6.5 can also be derived from Proposition 6.4. Indeed, if $z = Z_y x$, then $\mu_z \prec \mu_x$ by Proposition 6.4, while $\mu_z \prec \mu_y$ because of Proposition 6.2. Thus we see that, in any case, $\mu_z \prec \mu_x \wedge \mu_y$. If in addition $\mu_x \perp \mu_y$, then, of course, $\mu_z = 0$. But then $z = 0$, which shows that x is orthogonal to Z_y, and it follows at once that all of Z_x is too.

The following two theorems are of fundamental importance.

THEOREM 6.6. *Let E be a spectral measure on a Hilbert space H, and let $\{x_n\}$ be a finite or countably infinite sequence of vectors in H with the property that their measures μ_{x_n} are pairwise singular. Then the sum $\mathsf{Z} = \Sigma_n \oplus \mathsf{Z}_{x_n}$ (direct by virtue of Proposition 6.5) is also cyclic: $\mathsf{Z} = \mathsf{Z}_x$, where $\mu_x = \bigvee_n \mu_{x_n}$. A suitable generator x may be obtained by setting $x = \Sigma_n k_n x_n$ where the k_n's are any positive weights small enough so that the (orthogonal) series converges.*

PROOF. If x is constructed as stated, then it is clear that $x \in \mathsf{Z}$ and, because of the orthogonality of the various cycles Z_{x_n}, we have

$$\mu_x(M) = \left(E(M) \sum_n k_n x_n, \sum_n k_n x_n\right)$$

$$= \sum_n k_n^2 (E(M)x_n, x_n) = \sum_n k_n^2 \mu_{x_n}(M)$$

for every measurable set M. In other words, $\mu_x = \Sigma_n k_n^2 \mu_{x_n}$ which shows that $\mu_x = \bigvee_n \mu_{x_n}$, as stated. Hence to complete the proof it suffices to show that all of the vectors x_n belong to Z_x. To this end, fix an index n_0 and let $\nu = \Sigma_{n \neq n_0} k_n^2 \mu_{x_n}$. Then $\nu \perp \mu_{x_{n_0}}$, so there exists a measurable set M such that $\nu(M) = 0$ while $E(M)x_{n_0} = x_{n_0}$. But, since $\nu(M) = 0$, we have $\mu_{x_n}(M) = 0$, and therefore $E(M)x_n = 0$ for every $n \neq n_0$. Hence

$$E(M)x = \sum_n k_n E(M)x_n = k_{n_0} x_{n_0}$$

and the result follows. Q.E.D.

THEOREM 6.7. *Let E be a spectral measure on a Hilbert space H. Then the set of all measures μ_x, $x \in H$, is a σ-ideal J_E in $\mathfrak{M}$.*

PROOF. Let $\{x_n\}$ be a countable set of vectors, and suppose ν is a measure in $\mathfrak{M}$ such that $\nu \prec \bigvee_n \mu_{x_n}$. Then there exist pairwise singular measures $\nu_n \prec \mu_{x_n}$ such that $\nu = \bigvee_n \nu_n$ (disjointify!). Moreover, (Proposition 6.2) there exist vectors $y_n \in Z_{x_n}$ such that $\nu_n = \mu_{y_n}$ for every n. But then $\nu = \mu_z$ for a suitably chosen generator of the cycle $\Sigma_n \oplus Z_{y_n}$. Q.E.D.

Note. The σ-ideal J_E will be called the *σ-ideal of* E. Observe that Theorem 6.7 has the immediate corollary that, if K is any subspace of H that is invariant under E, then the set of measures μ_x, $x \in K$, is also a σ-ideal, for this is simply the σ-ideal of the spectral measure $E|K$.

7. **σ-ideals and subspaces.** Let E be a spectral measure from a measurable space $(\mathbf{M}, S)$ to H, and let J be a σ-ideal in the lattice $\mathfrak{M}$ of finite measures on $(\mathbf{M}, S)$. We shall write

$$F_J = \{x \in H : \mu_x \in J\}.$$

(If $J = (\mu)$, we write simply F_μ, instead of $F_{(\mu)}$.) *Trivial examples*: $F_{\mathfrak{M}} = H$, $F_0 = \{0\}$. For a more interesting class of examples, consider the case of the ideal (M) of measures supported on a fixed measurable set M: According to Lemma 6.3, a vector x belongs to $F_{(M)}$ if and only if $E(M)x = x$. In other words, $F_{(M)} = E(M)(H)$.

In all of these examples F_J turns out to be a subspace of H. As it happens, this holds true in general.

PROPOSITION 7.1. *F_J is a subspace of H.*

PROOF. If $E(M)x = E(M)y = 0$, then $E(M)(\alpha x + \beta y) = 0$ too. Thus, if z denotes a linear combination of x and y, then $\mu_z \prec \mu_x \vee \mu_y$, so that z belongs to F_J along with x and y. Hence F_J is a linear manifold. That F_J is also closed may be seen as follows. Suppose $\{x_n\}$ is a sequence in F_J such that $x_n \longrightarrow x$. Then $\mu_{x_n} \longrightarrow \mu_x$ setwise; whence it follows at once that $\mu_x \prec \bigvee_n \mu_{x_n}$, and therefore that $\mu_x \in J$. But then, of course, $x \in F_J$. Q.E.D.

Note. While it is true that $\mu_{x+y} \prec \mu_x + \mu_y$, it is *not* true in general that $\mu_{x+y} \leqslant \mu_x + \mu_y$. Indeed, direct computation shows that $\mu_{x+y} = \mu_x + 2 \operatorname{Re} \mu_{x,y} + \mu_y$, from which one derives (Cauchy-Schwarz inequality) the not very useful estimate

$$\mu_{x+y} \leqslant \mu_x + 2(\mu_x \mu_y)^{1/2} + \mu_y = (\mu_x^{1/2} + \mu_y^{1/2})^2.$$

Incidentally, we note once again, as in §6, that when $\mathcal{Z}_x \perp \mathcal{Z}_y$, i.e., when $\mu_{x,y} = 0$, the above expression for μ_{x+y} reduces to the equation $\mu_{x+y} = \mu_x + \mu_y$.

We shall write F_J for the projection of $\mathcal{H}$ onto the subspace $\mathcal{F}_J$, except that, if J is principal, $J = (\mu)$, we write F_μ instead of $F_{(\mu)}$. We observe, in passing, the obvious fact that, if J_E denotes the σ-ideal of E, then $\mathcal{F}_J = \mathcal{F}_{J \cap J_E}$. In other words, in employing this construction, one might as well stick to J's that are subideals of J_E. (In the finite-dimensional case *all* of the subideals of $J_E = (\Lambda)$ are simply the various ideals (M), $M \subset \Lambda$. A glance at Figure 1 shows how it was that Halmos [5] came to call the projections F_J the "columns" of E.)

The first use we make of Proposition 7.1 is to obtain an important complement to Theorem 6.7.

PROPOSITION 7.2. *If a system of vectors $\{x_\alpha\}$ has the property that the cycles $\mathcal{Z}_{x_\alpha}$ span $\mathcal{H}$, then the measures μ_{x_α} generate J_E. More generally, if $\{x_\alpha\}$ is an arbitrary collection of vectors in $\mathcal{H}$ then the measures μ_{x_α} generate the σ-ideal J consisting of the measures μ_x, $x \in \mathcal{K}$, where $\mathcal{K}$ denotes the smallest invariant subspace for E that contains all of the vectors x_α.*

PROOF. It suffices to prove the former assertion (cf. Note following Theorem 6.7). Let J denote the σ-ideal generated by the measures μ_{x_α}, and consider $\mathcal{F} = \mathcal{F}_J$. Since $\mathcal{Z}_{x_\alpha} \subset \mathcal{F}$ for all α, $\mathcal{F}$ contains a spanning set of vectors. Since $\mathcal{F}$ is a closed subspace, this implies that $\mathcal{F} = \mathcal{H}$, and the result follows. Q.E.D.

PROPOSITION 7.3. *The subspaces $\mathcal{F}_J$ and $\mathcal{F}_{J^\perp}$ are orthocomplements in $\mathcal{H}$ for every σ-ideal J.*

PROOF. Let x be a vector in $\mathcal{H}$, and let $\mu_x = \mu' \vee \mu''$, where $\mu' \in J$, $\mu'' \in J^\perp$. According to Theorem 2.1 there is an isomorphism of $\mathcal{Z}_x$ onto $L_2[\mu_x]$ that implements a unitary equivalence between E acting on $\mathcal{Z}_x$ and the standard spectral measure on $L_2[\mu_x]$ and carries x onto the function 1. Moreover, by Proposition 4.1 there exists a measurable set M such that $\mu' \sim \mu_x|M$ and $\mu'' \sim \mu_x|(\mathbf{M}\backslash M)$. It follows, of course, that the measures associated with the characteristic functions χ_M and $\chi_{(\mathbf{M}\backslash M)}$ by the standard spectral measure on $L_2[\mu_x]$ belong to J and $J^\perp$, respectively, and since $1 = \chi_M + \chi_{(\mathbf{M}\backslash M)}$, we conclude that there exist (unique) vectors y and z in $\mathcal{Z}_x$ such that $x = y + z$, with

$y \in \mathsf{F}_J$, $z \in \mathsf{F}_{J\perp}$. This shows that $\mathsf{H} = \mathsf{F}_J + \mathsf{F}_{J\perp}$ and, since we already know that F_J and $\mathsf{F}_{J\perp}$ are orthogonal (Proposition 6.5), the proof is complete. Q.E.D.

The great importance of the projections F_J in multiplicity theory stems from the property set forth in the next theorem. First, however, we need a geometrical lemma.

LEMMA 7.4. *Let M and N be orthogonal, isometric subspaces of a Hilbert space H, and let $\Phi: \mathsf{M} \longrightarrow \mathsf{N}$ be a unitary mapping of M onto N. Then there exists a unitary operator U on H such that $U|\mathsf{M} = \Phi$ and such that U commutes with every operator T on H having the properties that M and N are reducing subspaces for T and that $T\Phi(y) = \Phi(Ty)$ for all $y \in \mathsf{M}$.*

PROOF. Let L denote the subspace $(\mathsf{M} \oplus \mathsf{N})^{\perp}$, so that $\mathsf{H} = \mathsf{L} \oplus \mathsf{M} \oplus \mathsf{N}$. We may then simply define

$$U(x + y + z) = x + \Phi(y) + \Phi^*(z)$$

for all $x \in \mathsf{L}$, $y \in \mathsf{M}$, $z \in \mathsf{N}$, and straightforward calculation shows that U does all that is required of it. (It is rewarding to consider this construction matricially.) Q.E.D.

THEOREM 7.5. *Let E be a spectral measure on a Hilbert space H. Then the projections F_J associated with the various σ-ideals in $\mathfrak{M}$ (or, better, in J_E) are precisely the projections that doubly commute with E.*

PROOF. One half is easy: If T commutes with E, then $\mu_{Tx} \prec \mu_x$ for every vector x (Proposition 6.4), and consequently $Tx \in \mathsf{F}_J$ whenever $x \in \mathsf{F}_J$, no matter what σ-ideal J is. The hard part is to go the other way. Let F be a projection that commutes with every operator T that commutes with E and let $\mathsf{F} = F(\mathsf{H})$. If $F = F_J$ for some σ-ideal J, then it is clear that J must be the σ-ideal consisting of the various measures μ_x, $x \in \mathsf{F}$. (The latter is a σ-ideal, viz., the σ-ideal of the restriction $E|\mathsf{F}$ of E to F; see Note following Theorem 6.7.) Consequently, what is needed is to verify that, if y is an arbitrary vector in H with the property that there exists a sequence $\{x_n\}$ in F such that

$$\mu_y \prec \bigvee_n \mu_{x_n}, \tag{2}$$

then $Fy = y$.

Suppose, on the contrary, that (2) holds for some $y \notin \mathsf{F}$. Since $\mu_{(1-F)y} \prec \mu_y$ (for F commutes with E), we may and do assume that $y \perp \mathsf{F}$ (and $y \neq 0$). Moreover, it follows from Theorem 6.7 (and the following Note) that there exists a vector $z \in \mathsf{F}$ such that $\mu_y = \mu_z$. Now Z_y is orthogonal to F while $\mathsf{Z}_z \subset \mathsf{F}$. Hence, in particular, $\mathsf{Z}_y \perp \mathsf{Z}_z$. But the restrictions of E to these two invariant subspaces are unitarily equivalent (for they are both unitarily equivalent to the standard spectral measure on $L_2[\mu_y = \mu_z]$).

Hence, by the preceding lemma, there exists a unitary operator U on $\mathcal{H}$ that commutes with E and that carries $\mathcal{Z}_x$ onto $\mathcal{Z}_y$. But then $\mathcal{F}$ is not invariant under U, and consequently F does not commute with U, contrary to hypothesis. Q.E.D.

8. Multiplicity: The solution. The broad outline of the sort of decomposition we are after is beginning to take shape; we want to break $\mathcal{H}$ up into pieces, each of which is a direct sum of the form $\Sigma_\alpha \oplus \mathcal{Z}_{x_\alpha}$ where all of the measures μ_{x_α} are mutually equivalent. A subspace that can be written as such a direct sum will be called by the suggestive, if inelegant, name *stack*; and if all of the measures μ_{x_α} are equivalent to μ, then we shall refer to it as a *stack with measure* μ. If the cardinal number of cycles $\mathcal{Z}_{x_\alpha}$ in a stack is c, then the stack is *c-fold*, or has *multiplicity* c. The following lemma is the essential ingredient of the main uniqueness theorem.

LEMMA 8.1. *Let μ be a nonzero measure belonging to $\mathfrak{M}$, and let $\mathcal{H}$ denote the direct sum of n copies of $L_2[\mu]$, where n denotes a positive integer. Moreover, let E denote the standard spectral measure on $\mathcal{H}$ obtained by ampliating the standard spectral measure on $L_2[\mu]$:*

$$E(M)\{f_1, \cdots, f_n\} = \{\chi_M f_1, \cdots, \chi_M f_n\}.$$

Suppose given a system $F_1, \cdots, F_m$ of elements of $\mathcal{H}$ with the properties that their cyclic subspaces $\mathcal{Z}_{F_i}$ are pairwise orthogonal and that each measure μ_{F_i} is equivalent to μ; then $m \leqslant n$.

PROOF. We may and do assume that μ_{F_i} is actually equal to μ for all i, since we need only replace each F_i by an equivalent element in $\mathcal{Z}_{F_i}$ to arrange for this to be so. Let $F_i = \{f_1^i, \cdots, f_n^i\}$, $i = 1, \cdots, m$. Then

$$\mu(M) = \mu_{F_i}(M) = \sum_{j=1}^{m} \|\chi_M f_j^i\|^2 = \sum_{j=1}^{n} \int_M |f_j^i|^2 \, d\mu = \int_M \left[\sum_{j=1}^{n} |f_j^i|^2\right] d\mu$$

for every measurable set M, which implies that

$$\sum_{j=1}^{n} |f_j^i|^2 \equiv 1 \ [\mu], \quad i = 1, \cdots, m.$$

The other hypothesis we are given is that the cycles $\mathcal{Z}_{F_i}$ are pairwise orthogonal, which implies that $\{\chi_M f_1^i, \chi_M f_2^i, \cdots, \chi_M f_n^i\}$ is orthogonal to $\{f_1^j, f_2^j, \cdots, f_n^j\}$ for every measurable set M whenever $i \neq j$. But then

$$\sum_{k=1}^{n} \int_M f_k^i \bar{f}_k^j \, d\mu = \int_M \left[\sum_{k=1}^{n} f_k^i \bar{f}_k^j\right] d\mu = 0$$

for every measurable set M whenever $i \neq j$. But then

$$\sum_{k=1}^{n} f_k^i \bar{f}_k^j \equiv 0 \ [\mu], \qquad i \neq j.$$

This shows that the n-tuples $(f_1^i(\lambda), \cdots, f_n^i(\lambda))$, $i = 1, \cdots, m$, form an orthonormal set in $\mathbf{C}^n$ for almost every $\lambda \in \mathbf{M}$, and since $\mu \neq 0$, this assures that they form an orthonormal set at at least *one* point. Hence $m \leqslant n$. Q.E.D.

Note. The lemma is, in fact, trivially false for $\mu = 0$.

THEOREM 8.2. *Let E be a spectral measure on a Hilbert space $\mathcal{H}$, let μ be a nonzero measure belonging to $\mathfrak{M}$, and suppose $\mathcal{S}$ and $\mathcal{T}$ are c- and c'-fold stacks, respectively, with respect to E, such that $\mathcal{S}$ and $\mathcal{T}$ both have measure μ and such that $\mathcal{S} \subset \mathcal{T}$. Then $c \leqslant c'$. In particular, if $\mathcal{S}$ is expressed in any other way as a stack, then there must be exactly c cycles employed in the expression.*

PROOF. If c is finite, then the only case that requires investigation is when c' is finite too, and this case is precisely the one covered by the lemma, since it is clear that the restriction of E to a stack such as $\mathcal{T}$ is unitarily equivalent (in more or less unique fashion) to the standard spectral measure on the direct sum of c' copies of $L_2[\mu]$. If $c = \aleph_0$, then by the finite case we must have $c' \geqslant m$ for every positive integer, and therefore $c' \geqslant \aleph_0$. Finally if $c > \aleph_0$, then $\dim \mathcal{S} = c \cdot \aleph_0 = c$, and since this implies $\dim \mathcal{T} \geqslant c$, it follows that $\dim \mathcal{T} = c'$, and hence that $c' \geqslant c$. Finally, to verify the second assertion of the theorem, we observe that if $\mathcal{S}$ is expressed as a stack in one way with measure μ and in another way with measure ν, then $\mu \sim \nu$ since $(\mu) = (\nu)$ is the σ-ideal of the restriction of E to $\mathcal{S}$. Hence the second assertion follows at once from the first. Q.E.D.

Before stating the main theorem, it is desirable to establish just one last lemma; the result is almost trivial, but it would be untidy to repeat this construction over and over.

LEMMA 8.3. *If $\mathcal{S}$ is a c-fold stack with measure μ, and if $\nu \prec \mu$, then $\mathcal{S} \cap \mathcal{F}_\nu$ (which is clearly the same thing as $\mathcal{F}_\nu$ for the restriction $E|\mathcal{S}$) is a c-fold stack with measure ν.*

PROOF. Suppose $\mathcal{S} = \Sigma_{\alpha \in A} \oplus \mathcal{Z}_{x_\alpha}$ where $\mu_{x_\alpha} \sim \mu$ for all α and card $A = c$. In each $\mathcal{Z}_{x_\alpha}$ there is a unique cycle $\mathcal{Z}_{y_\alpha}$ such that $\mu_{y_\alpha} \sim \nu$. (The vector y_α is not unique, of course, but the cycle is.) It is clear that $\Sigma_\alpha \oplus \mathcal{Z}_{y_\alpha}$ is a c-fold stack with measure ν. Since $\mathcal{Z}_{y_\alpha}$ consists precisely of those vectors in $\mathcal{Z}_{x_\alpha}$ that belong to $\mathcal{F}_\nu$, it is clear that all that is needed is to show that a sum $u = \Sigma_\alpha u_\alpha$, $u_\alpha \in \mathcal{Z}_{x_\alpha}$ for all α, belongs to $\mathcal{F}_\nu$ if and only if each u_α does, and that is more or less self-evident. Q.E.D.

THEOREM 8.4. *Let E be a spectral measure on a Hilbert space $\mathcal{H}$. Then there exist pairwise singular, nonzero measures $\{\mu_\alpha\}$ and corresponding subspaces S_α such that each S_α is a stack with measure μ_α and such that $\mathcal{H}$ is the direct sum of the S_α's: $\mathcal{H} = \Sigma_\alpha \oplus S_\alpha$. This decomposition is unique in the following sense. Suppose $\{\nu_\beta\}$ and $\{T_\beta\}$ is another such system of measures and stacks for E. For each cardinal number c write A_c for the set of those indices α such that S_α has multiplicity c, and, similarly, write B_c for the set of indices β such that T_β has multiplicity c. Then for each c the σ-ideal generated by the set of measures $\{\mu_\alpha\colon \alpha \in A_c\}$ coincides with that generated by the set $\{\nu_\beta\colon \beta \in B_c\}$.*

PROOF OF EXISTENCE. Start with any measure $\mu \neq 0$ in $\mathcal{J}_E$. (If $\mathcal{H} = \{0\}$, then the empty collection of stacks satisfies the conditions of the theorem.) By Zorn's lemma there exists a (nonempty) maximal family of orthogonal cycles $\mathcal{Z}_x$ such that $\mu_x \sim \mu$ for every x. Choose any one such maximal family and denote its direct sum by S. Let G denote the orthogonal difference $G = F_\mu \ominus S$. If $G = \{0\}$, then $S = F_\mu$. Otherwise the various measures μ_u, $u \in G$, form a σ-ideal contained in (μ), and therefore a principal ideal, say (ν). Then $0 \neq \nu \prec \mu$. Moreover $\nu \not\sim \mu$, for otherwise G would contain another cycle with measure μ, which would contract the maximality of S. Hence, if we write

$$\mu = \nu \vee \nu', \qquad \nu \perp \nu',$$

then $\nu' \neq 0$. But then by Lemma 8.3, $S' = S \cap F_{\nu'}$ is a stack with measure ν', and, as is easily seen, $S' = F_{\nu'}$. Thus we have proved the following partial result: If $\mathcal{H} \neq \{0\}$, then $\mathcal{H}$ contains a stack $S \neq \{0\}$ with some measure $\nu \neq 0$ and some multiplicity $c \geqslant 1$ such that $S = F_\nu$.

Now by Zorn's lemma, once again, there exists an orthogonal family of stacks $\{S_\alpha\}$ which is maximal with respect to the property that, if μ_α is the measure of S_α, then $S_\alpha = F_{\mu_\alpha}$. Let $\mathcal{J}$ denote the σ-ideal generated by all of the various measures μ_α. It is clear that the subspace $F_{\mathcal{J}}$ coincides with the direct sum $\Sigma_\alpha \oplus S_\alpha$, and hence that $K = \mathcal{H} \ominus \Sigma_\alpha \oplus S_\alpha = F_{\mathcal{J}^\perp}$, where $\mathcal{J}^\perp$ denotes, as usual, the complement of $\mathcal{J}$ in $\mathfrak{M}$ (or, equally well, in $\mathcal{J}_E$). Suppose $K \neq \{0\}$. Then, as we have just seen, K contains a stack $S \neq \{0\}$ with some measure $\nu \neq 0$ such that S coincides with F_ν for the spectral measure $E|K$. But since $\nu = \mu_x$ for some vector $x \in K$, and $K = F_{\mathcal{J}^\perp}$, the measure ν belongs to $\mathcal{J}^\perp$, and it follows that F_ν is the same, whether computed for E or $E|K$. In other words, S is a stack with measure $\nu \neq 0$ that has the property $S = F_\nu$, and $S \perp S_\alpha$ for every index α. Since this contradicts the maximality of $\{S_\alpha\}$, we conclude that $K = \{0\}$, and hence that $\mathcal{H} = \Sigma_\alpha \oplus S_\alpha$.

To complete the proof we have only to observe that if $\alpha \neq \beta$ then, by construction, $\mathcal{F}_{\mu_\alpha}$ and $\mathcal{F}_{\mu_\beta}$ are orthogonal, so that, necessarily, $\mu_\alpha \perp \mu_\beta$.

PROOF OF UNIQUENESS. By symmetry it suffices to show that if $\alpha \in A_c$, then μ_α belongs to the σ-ideal generated by $\{\nu_\beta\colon \beta \in B_c\}$. Let β be any index such that μ_α and ν_β are not mutually singular, and let $\nu = \mu_\alpha \wedge \nu_\beta$. Consider the subspace $\mathcal{F}_\nu$. Since ν is singular to every $\mu_{\alpha'}$, $\alpha' \neq \alpha$, it is clear that $\mathcal{F}_\nu \subset \mathcal{S}_\alpha$. Similarly, $\mathcal{F}_\nu \subset \mathcal{T}_\beta$. But then $\mathcal{F}_\nu = \mathcal{F}_\nu \cap \mathcal{S}_\alpha$ is, by Lemma 8.3, a stack with measure ν and multiplicity c while, also, $\mathcal{F}_\nu = \mathcal{F}_\nu \cap \mathcal{T}_\beta$ is a stack with measure ν and multiplicity equal to that of $\mathcal{T}_\beta$. Consequently, by Theorem 8.2, $\beta \in B_c$. In other words, μ_α is singular with respect to every measure ν_β such that β is not in B_c. But μ_α is certainly in the σ-ideal generated by all of the measures ν_β, for the latter coincides with $\mathcal{J}_E$. Hence the result follows by distributivity. Q.E.D.

NOTATION AND DEFINITION. The σ-ideal generated by the various measures of stacks of multiplicity c in one (and hence in every) decomposition of the type described in Theorem 8.4 will be denoted by $\mathcal{J}_c$. The assignment $c \longrightarrow \mathcal{J}_c$ is the *multiplicity function* of E. (We use only c's that are $\geqslant 1$; there is no need to use c's that are greater than dim $\mathcal{H}$ since $\mathcal{J}_c = 0$ for all such cardinal numbers.)

Note. The multiplicity function defined in §3 in the finite-dimensional case is not quite the same as the one defined here. There we wrote $m(c) = \Lambda_c$; here we are now proposing the emendation $m(c) = (\Lambda_c)$.

THEOREM 8.5. *A necessary and sufficient condition for two spectral measures E and F (on the same measurable space) to be unitarily equivalent is that they possess identical multiplicity functions.*

PROOF. If Φ is a unitary mapping such that $\Phi^* E \Phi = F$, and if $\mathcal{S}$ is a stack for F with measure μ and multiplicity c, and with the property that $\mathcal{S} = \mathcal{F}_\mu$, then it is clear that $\Phi(\mathcal{S})$ is a stack for E with the same measure and multiplicity, and that $\Phi(\mathcal{S})$ is also equal to $\mathcal{F}_\mu$. It follows, of course, that any decomposition of F as in Theorem 8.4 is carried by Φ onto a similar decomposition of E, and hence that E and F possess the same multiplicity function.

Suppose now, conversely, that the multiplicity functions of E and F coincide. Let c denote a cardinal number ($\geqslant 1$) for which $\mathcal{J}_c \neq 0$ (for both E and F). It will suffice to show that $E|\mathcal{F}_{\mathcal{J}_c}$ and $F|\mathcal{F}_{\mathcal{J}_c}$ are unitarily equivalent, for we can then simply form the direct sum of the various unitary mappings. In other words, we may as well assume that $\mathcal{J}_E = \mathcal{J}_F = \mathcal{J}_c = \mathcal{J}$. Then the Hilbert space $\mathcal{H}$ on which E acts is the direct sum $\mathcal{H} = \Sigma_\alpha \oplus \mathcal{S}_\alpha$ of a family of stacks, each of multiplicity c, having measures μ_α that are pairwise singular and that together generate $\mathcal{J}$.

Let $I = \{\iota\}$ be an index family of cardinal number c, and in each S_α choose vectors $x_{\alpha,\iota}$ such that $S_\alpha = \Sigma_\iota \oplus Z_{x_{\alpha,\iota}}$ where $\mu_{x_{\alpha,\iota}} = \mu_\alpha$ for every ι. Let $L_2^c[\mu_\alpha]$ denote the direct sum of a family of copies of $L_2[\mu_\alpha]$ indexed by I. Then (Theorem 2.1) there is a unique unitary mapping of S_α onto $L_2^c[\mu_\alpha]$ that carries each $x_{\alpha,\iota}$ onto the function 1 in the ιth copy of $L_2[\mu_\alpha]$, and this mapping carries the restriction of E to S_α onto the *standard* spectral measure on $L_2^c[\mu_\alpha]$ (by which is meant, of course, the ampliation to $L_2^c[\mu_\alpha]$ of the standard spectral measure on $L_2[\mu_\alpha]$; cf. Lemma 8.1). Thus, altogether, E is unitarily equivalent to the direct sum—call it $\hat{E}$—of the various standard spectral measures on the spaces $L_2^c[\mu_\alpha]$ acting on the space $\mathsf{L} = \Sigma_\alpha \oplus L_2^c[\mu_\alpha]$.

Now let L_0 denote the direct sum $\Sigma_\alpha \oplus L_2[\mu_\alpha]$ of the spaces $L_2[\mu_\alpha]$ *without* any replication, and let $\widetilde{E}$ denote the spectral measure on L_0 obtained by forming the direct sum of the various standard spectral measures on the spaces $L_2[\mu_\alpha]$. Then L may be identified in obvious fashion with the direct sum of c copies of L_0, and under this identification $\hat{E}$ is identified with the ampliation on L of the spectral measure $\widetilde{E}$. Similarly, of course, the spectral measure F is unitarily equivalent to the c-fold ampliation of the spectral measure $\widetilde{F}$ obtained by forming the direct sum of the standard spectral measures on some family of spaces $L_2[\nu_\beta]$, where the family of measures $\{\nu_\beta\}$ is pairwise singular and generates the σ-ideal $\mathcal{J}$. Thus we see that it suffices to prove the following multiplicity-free lemma.

LEMMA 8.6. *Suppose $\{\mu_\alpha\}$ and $\{\nu_\beta\}$ are two systems of pairwise singular, nonzero measures that generate the same σ-ideal. Let $\mathsf{L}_0 = \Sigma_\alpha \oplus L_2[\mu_\alpha]$, let $\mathsf{M}_0 = \Sigma_\beta \oplus L_2[\nu_\beta]$, and let E_0 and F_0, respectively, denote the spectral measures on L_0 and M_0 obtained by forming the direct sums of the various standard spectral measures on the spaces $L_2[\mu_\alpha]$ and $L_2[\nu_\beta]$. Then E_0 and F_0 are unitarily equivalent.*

PROOF. Let $\xi_{\alpha,\beta} = \mu_\alpha \wedge \nu_\beta$. Then $\mu_\alpha = \bigvee_\beta \xi_{\alpha,\beta}$ and, since the measures ν_β are pairwise singular, it follows that $L_2[\mu_\alpha]$ may be identified with the direct sum $\Sigma_\beta \oplus L_2[\xi_{\alpha,\beta}]$. Moreover, under this identification, which amounts to nothing more than breaking **M** up into (almost) disjoint pieces, the standard spectral measure on $L_2[\mu_\alpha]$ is identified with the direct sum of the standard spectral measures on the spaces $L_2[\xi_{\alpha,\beta}]$. Hence E_0 is unitarily equivalent to the direct sum of all the standard spectral measures on the spaces $L_2[\xi_{\alpha,\beta}]$. Similarly, F_0 is also unitarily equivalent to the direct sum of all the standard spectral measures on the spaces $L_2[\xi_{\alpha,\beta}]$. Q.E.D.

COROLLARY 8.7. *A necessary and sufficient condition for two normal operators A and B to be unitarily equivalent is that their spectral measures should possess identical multiplicity functions.*

9. Uniform multiplicity one. A spectral measure E is said to be of *uniform multiplicity* c if $m(c) = J_E$ for some cardinal number c. It is the substance of Theorem 8.4 that every spectral measure is the direct sum in a unique way of pieces of different uniform multiplicities, and that each of the latter is simply an ampliation (not in a unique manner) of a spectral measure of uniform multiplicity one. It would seem appropriate, therefore, to give special attention to this latter case, and in this connection we have the following result.

THEOREM 9.1. *For any spectral measure E on a Hilbert space $\mathcal{H}$ the following are equivalent:*

(i) *E has uniform multiplicity one.*

(ii) *There do not exist orthogonal, nontrivial reducing subspaces $\mathcal{M}$ and $\mathcal{N}$ for E such that $E|\mathcal{M}$ and $E|\mathcal{N}$ are unitarily equivalent.*

(iii) *$\mathcal{Z}_x = \mathcal{F}_{\mu_x}$ for every vector $x \in \mathcal{H}$.*

(iv) *If $\mathcal{Z}_x$ and $\mathcal{Z}_y$ are orthogonal cycles, then $\mu_x \perp \mu_y$.*

(v) *If F is any projection that commutes with E, then F doubly commutes with E.*

PROOF. (i) $\Rightarrow$ (ii) Suppose $\mathcal{M}$ and $\mathcal{N}$ are orthogonal reducing subspaces and that $E|\mathcal{M}$ and $E|\mathcal{N}$ are unitarily equivalent via Φ. Let $x \in \mathcal{M}$ and set $y = \Phi(x)$. Let $\mathcal{H} = \Sigma_\alpha \oplus \mathcal{Z}_\alpha$ be a decomposition of $\mathcal{H}$ as in Theorem 8.4, so that the measures μ_α of the $\mathcal{Z}_\alpha$'s (considered as stacks of multiplicity one) are pairwise singular, and suppose $\nu = \mu_x \wedge \mu_\alpha \neq 0$ for some index α. Then the cycle $\mathcal{F}_\nu \subset \mathcal{Z}_\alpha$ contains an orthogonal pair of cycles $\mathcal{Z}_{x'} \subset \mathcal{Z}_x$ and $\mathcal{Z}_{y'} \subset \mathcal{Z}_y$ with $\mu_{x'} = \mu_{y'} = \nu$, a clear cut violation of Theorem 8.2. Thus we must have $\mu_x \perp \mu_\alpha$ for every α, so that $\mu_x = 0$. But then $x = 0$, and it follows that $\mathcal{M} = \mathcal{N} = \{0\}$.

(ii) $\Rightarrow$ (iii) Suppose, on the contrary, that $\mathcal{Z}_x \neq \mathcal{F}_{\mu_x}$ for some vector x, and hence that there exists a vector $y \notin \mathcal{Z}_x$ such that $\mu_y \prec \mu_x$. Replacing y by its projection y' on the orthocomplement of $\mathcal{Z}_x$, we obtain a vector y' such that $\mathcal{Z}_{y'} \perp \mathcal{Z}_x$ and such that $\mu_{y'} \prec \mu_x$. Next, replacing x by a suitable vector $x' \in \mathcal{Z}_x$, we obtain a pair of vectors x', y' such that $\mathcal{Z}_{x'} \perp \mathcal{Z}_{y'}$ while $\mu_{x'} = \mu_{y'}$. But then $E|\mathcal{Z}_{x'}$ and $E|\mathcal{Z}_{y'}$ are both unitarily equivalent to the same standard spectral measure, contrary to (ii).

(iii) $\Rightarrow$ (iv) If $\mathcal{Z}_x \perp \mathcal{Z}_y$ and if (iii) holds, then, of course, $\mathcal{F}_{\mu_x} \perp \mathcal{F}_{\mu_y}$, and it follows at once that $\mu_x \perp \mu_y$.

(iv) $\Rightarrow$ (v) Suppose $\mathcal{F}$ is a subspace invariant under E, and let $\mathcal{J}$ denote the σ-ideal of measures μ_x, $x \in \mathcal{F}$. According to Theorem 7.4, what we must show is that $\mathcal{F} = \mathcal{F}_{\mathcal{J}}$. Suppose, on the contrary, that there exists a vector $y \in \mathcal{F}_{\mathcal{J}} \backslash \mathcal{F}$. Then $(1 - F)y = y' \neq 0$, where F denotes the projection of $\mathcal{H}$ onto $\mathcal{F}$, and $y' \perp \mathcal{F}$, so that if $x \in \mathcal{F}$, then $Z_{y'} \perp Z_x$, and therefore, by (iv), $\mu_{y'} \perp \mu_x$. This shows that $\mu_{y'} \perp \mathcal{J}$, and hence that $y' \notin \mathcal{F}_{\mathcal{J}}$. But $\mathcal{F}_{\mathcal{J}}$ is reducing for F since, as we assume, F commutes with E. Hence y' must belong to $\mathcal{F}_{\mathcal{J}}$ along with y.

(v) $\Rightarrow$ (i) Let $\mathcal{H} = \Sigma_\alpha \oplus S_\alpha$ be a decomposition of $\mathcal{H}$ into stacks of uniform multiplicity, according to Theorem 8.4, and suppose that one of the stacks S_α (with measure $\mu_\alpha \neq 0$) has multiplicity at least two. Then, if x and y are any two vectors in S_α such that $\mu_x = \mu_y = \mu_\alpha$ while $Z_x \perp Z_y$, there exists a unitary operator U that commutes with E and interchanges Z_x and Z_y (recall the proof of Theorem 7.4), and while the cyclic projections Z_x and Z_y do commute with E they clearly do not commute with U. Q.E.D.

10. The separable case. If the Hilbert space $\mathcal{H}$ is separable, then things are a little simpler. In the first place, the σ-ideal $\mathcal{J}_E$ is countably generated (Proposition 7.2) and is therefore, in fact, principal. Thus there exists a single finite measure, call it μ_E (unique only up to equivalence, of course) such that $\mathcal{J}_E = (\mu_E)$. Likewise, all of the σ-ideals appearing in the definition of the multiplicity function are also principal, so that in the separable case it is possible, and is indeed customary, to define the multiplicity function in such a way that $m(c)$ is a measure μ_c rather than the σ-ideal (μ_c). When this is done, Theorem 8.4 assumes the following form.

THEOREM 8.4$'$. *Let E be a spectral measure on a separable Hilbert space $\mathcal{H}$. Then there exist pairwise singular measures μ_n, and corresponding subspaces S_n, $n = 1, 2, \cdots, \aleph_0$, such that each S_n is a stack of multiplicity n having measure μ_n, and such that $\mathcal{H}$ is the direct sum of the S_n's:*

$$\mathcal{H} = \sum_{n=1}^{\aleph_0} \oplus S_n.$$

In this decomposition the subspaces S_n and the multiplicities are absolutely unique, while the measures μ_n are unique up to equivalence.

Note. While this way of wording Theorem 8.4 has the advantage that it obviates the necessity of introducing σ-ideals, it introduces a small amount of arbitrariness. It is, after all, precisely the *ideals* (μ_n) that are uniquely determined by E, not the measures μ_n. Thus there is something to be said for Wecken and his σ-ideals [**13**], even when H is separable. Incidentally, it should not go unnoticed that the measure μ_E mentioned above has the property that it is *equivalent to* E itself, in the sense that $E(M) = 0$ if and only if $\mu_E(M) = 0$.

When H is separable, there is still another condition that can be added to the list in Theorem 9.1.

THEOREM 9.1′. *Let E be a spectral measure on a separable Hilbert space H. Then the five conditions of Theorem 9.1 are all equivalent to the following:*

(vi′) *There exists a cyclic vector for E.*

PROOF. If x_0 is cyclic for E, then, as we have known since Theorem 2.1, E is unitarily equivalent to the standard spectral measure on $L_2[\mu_{x_0}]$, and therefore clearly has uniform multiplicity one. This part of the argument is obviously valid in general, and has nothing to do with separability. Suppose on the other hand that E has uniform multiplicity one. Then in Theorem 8.4′ we have $m(1) = \mu_E$ and $S_1 = H$. Thus E is unitarily equivalent to the standard spectral measure on $L_2[\mu_E]$, and clearly possesses a cyclic vector. Q.E.D.

There is yet another simplification that turns up whenever the ideal J_E is principal. Indeed, in this case it is not only true that the σ-ideals J contained in J_E are principal; more to the point, each such J is actually of the form $(\mu_E|M)$ for some measurable set M. Consequently, $J = (\mu_E) \cap (M)$ (recall the opening paragraph of §7). But then $F_J = F_{(M)}$, and therefore $F_J = E(M)$. We have proved the following interesting result.

THEOREM 10.1 *If E is a spectral measure on a Hilbert space H, and if the σ-ideal J_E of E is principal, then the only projections that doubly commute with E are the spectral projections $E(M)$ themselves. In particular this is true when H is separable.*

COROLLARY 10.2. *Let A be a normal operator on a separable Hilbert space H. Then the only operators in $L(H)$ that doubly commute with A (and A^*) are the various operators of the form $f(A)$.*

PROOF. Suppose T commutes with every operator that commutes with A (and A^*). Then, in particular, T^* commutes with A (and A^*) so T is normal, and therefore possesses a spectral measure E_T. The spectral projections $E_T(M)$ commute with everything that T does, so they too doubly commute with A (and A^*). Hence the range of E_T is contained in that of E_A. But then it is easily seen that T is in fact a function $\int f dE_A$ of A. (Approximate the identity function λ on the spectrum of T by simple functions in a suitably neat manner.)

Finally, because of its historical interest, we indicate an alternate scheme for defining a multiplicity function that works well in the separable case.

THEOREM 10.3 (HELLINGER [6]). *Let E be a spectral measure on a separable Hilbert space $\mathcal{H}$. Then there exists a (finite or infinite) sequence $\{\nu_n\}$ of nonzero measures such that*

$$\nu_1 \succ \nu_2 \succ \cdots \succ \nu_n \succ \cdots$$

and a corresponding sequence $\{y_n\}$ of vectors such that $\mu_{y_n} = \nu_n$ for every n and such that $\mathcal{H} = \Sigma_n \oplus \mathcal{Z}_{y_n}$. The measures ν_n are uniquely determined by E up to equivalence.

PROOF OF EXISTENCE. Let $m(n) = \mu_n$ be the multiplicity function of E and let $\mathcal{H} = \Sigma_n \oplus \mathcal{S}_n$ be the decomposition of Theorem 8.4′. For each positive integer k set

$$\nu_k = \bigvee_{n=k}^{\aleph_0} \mu_n.$$

Then it is clear that $\nu_1 \succ \nu_2 \succ \cdots$. (If some $\nu_n = 0$, then we stop and keep only the nonzero ν_k's.) In each stack $\mathcal{S}_n$ choose a vector x_n such that $\mu_{x_n} = \mu_n$, and form the sum $\mathcal{Z} = \Sigma_n \oplus \mathcal{Z}_{x_n}$. Since the measures μ_n are pairwise singular, $\mathcal{Z}$ is also a cycle (Theorem 6.6). Choose a vector $y_1 \in \mathcal{Z}$ such that $\mathcal{Z} = \mathcal{Z}_{y_1}$. Then $\mu_{y_1} = \nu_1$, and the multiplicity function m_1 of the spectral measure $E|(\mathcal{H} \ominus \mathcal{Z}_y)$ is clearly given by

$$m_1(n) = \mu_{n+1}, \qquad n = 1, 2, \cdots, \aleph_0.$$

The balance of the obvious induction is left to the reader.

PROOF OF UNIQUENESS. Set $\mathcal{S}_n = \mathcal{F}_{\nu_n} \ominus \mathcal{F}_{\nu_{n+1}}$, $n = 1, 2, \cdots$. (Since $\mathcal{F}_{\nu_1} = \mathcal{H}$, we have, in particular, $\mathcal{S}_1 = \mathcal{H} \ominus \mathcal{F}_{\nu_2}$.) It is easily seen that each $\mathcal{S}_n$ is a stack of multiplicity n, and it is obvious that the measures of the various $\mathcal{S}_n$'s are pairwise singular. Hence, if $\mathcal{H} = \Sigma_n \oplus \mathcal{S}_n$, then we have here the unique decomposition of Theorem 8.4′. On the other hand, if $\mathcal{K} = \Sigma_n \oplus \mathcal{S}_n \neq \mathcal{H}$, and if $x \in \mathcal{H} \ominus \mathcal{K}$, then $\mu_x < \nu_n$ for every n. Conversely, if $\mu_x \prec \nu_n$ for every n, then $x \in \mathcal{F}_{\nu_n}$ for every n and therefore $x \perp \mathcal{K}$. Thus $\mathcal{H} \ominus \mathcal{K} = \mathcal{F}_\nu$ where $\nu = \bigwedge_n \nu_n$. But the space $\mathcal{F}_\nu = \mathcal{S}_{\aleph_0}$ is clearly itself a stack with measure ν and multiplicity $\aleph_0$. Hence, in this case,

$$\mathcal{H} = \sum_{n=1}^{\aleph_0} \oplus \mathcal{S}_n$$

is the unique decomposition of Theorem 8.4′. Since in either case the measures

μ_n of the various stacks S_n are unique up to equivalence, and since the measures ν_n can be recaptured as

$$\nu_n = \bigvee_{k \geqslant n} \mu_k,$$

it is clear that the ν_n's also are unique up to equivalence. Q.E.D.

11. The L_2-space of a σ-ideal.[1] The only significant difference that arises in passing from the separable to the nonseparable case as far as multiplicity theory is concerned is that in the nonseparable case one *must* employ σ-ideals of measures instead of the measures themselves. As a result, there are in the general case no obvious candidates to serve as "standard" building blocks as the standard spectral measures do in the separable case. We conclude this note with an indication of a way of associating a standard spectral measure with a σ-ideal of measures. For those in the know it would probably suffice to say "form the obvious direct limit." For those not in the know, we introduce the appropriate abstract paraphernalia.

Suppose D is a directed set, and suppose we are given (i) for each index $\lambda \in D$ a linear space E_λ, and (ii) for each pair of indices λ, λ' such that $\lambda \leqslant \lambda'$ a linear isomorphism $\phi_\lambda^{\lambda'}: E_\lambda \to E_{\lambda'}$ of E_λ into $E_{\lambda'}$. Suppose further that these mappings are *coherent* in the sense that

$$\phi_\lambda^{\lambda''} = \phi_{\lambda'}^{\lambda''} \circ \phi_\lambda^{\lambda'}$$

whenever $\lambda \leqslant \lambda' \leqslant \lambda''$. If $x \in E_\lambda$ and $y \in E_{\lambda'}$, we write $x \sim y$ if and only if there exists $\lambda'' \geqslant \lambda, \lambda'$ such that $\phi_\lambda^{\lambda''}(x) = \phi_{\lambda'}^{\lambda''}(y)$. A simple verification shows that $\sim$ is an equivalence relation on the union $\bigcup_\lambda E_\lambda$, and, because the mappings $\phi_\lambda^{\lambda'}$ are all one-to-one, no equivalence class $[x]$ possesses more than one representative in any one space E_λ. Moreover, if $x \sim x'$ and $y \sim y'$, and if E_λ contains both x and y and $E_{\lambda'}$ contains both x' and y', then it is clear that $\alpha x + \beta y \sim \alpha x' + \beta y'$, so that we may and do define

$$\alpha[x] + \beta[y] = [\alpha x + \beta y].$$

The resulting linear space, call it E_D, is known as the *direct limit* of the family $\{E_\lambda\}_{\lambda \in D}$ (with respect to the mappings $\phi_\lambda^{\lambda'}$). It is a triviality to verify that if the spaces E_λ are all *Hilbert* spaces, and if the mappings $\phi_\lambda^{\lambda'}$ are all isometries, and if we write

$$([x], [y]) = (x, y)_\lambda,$$

[1] For more general treatment of this topic, the reader is referred to [12].

where $(\, , \,)_\lambda$ denotes the inner product in E_λ, and λ is any index large enough so that E_λ contains representatives of both $[x]$ and $[y]$, then $(\, , \,)$ so defined on E_D turns E_D into an inner product space. In general, of course, there is no reason to expect this space to be complete, and one passes to the completion in order to arrive at the appropriate notion of direct limit for Hilbert spaces. In the application we have in mind here, however, the space E_D will turn out to be complete already.

PROPOSITION 11.1. *Let J be a σ-ideal of measures in $\mathfrak{M}$. For each $\mu \in J$ let $H_\mu = L_2[\mu]$, and for each pair of measures μ, ν such that $\mu \prec \nu$ define*

$$\phi_\mu^\nu(f) = \psi^{1/2} f, \quad f \in H_\mu,$$

where $\psi = d\mu/d\nu$ denotes the Radon-Nikodym derivative of μ with respect to ν. (The function ψ is only defined up to sets of measure zero $[\nu]$, but the mapping ϕ_μ^ν is uniquely determined by μ and ν.) Then the family $\{\phi_\mu^\nu\}$ is a coherent system of isometries, and the direct limit of the family $\{H_\mu\}_{\mu \in J}$ is a (complete) Hilbert space.

PROOF. It is routine to verify that the mappings ϕ_μ^ν are determined by the measures μ and ν, i.e., do not depend on the particular choice of ψ, and that they form a coherent system of isometries. (It is a part of the Radon-Nikodym theorem that if $\lambda \prec \mu \prec \nu$, then $d\lambda/d\nu = (d\lambda/d\mu)(d\mu/d\nu)$.) The only other thing that requires proof is the completeness of the space H_J, and this follows at once from the fact that J is σ-complete for this insures that if $\{[f]_n\}$ is a Cauchy sequence of elements of H_J, then there exists a single measure μ such that all of the equivalence classes $[f]_n$ are simultaneously represented in H_μ. Q.E.D.

Let us denote the Hilbert space of Proposition 11.1 by $L_2[J]$. If M is a measurable set, and if f belongs to $L_2[\mu]$, then it is clear that

$$\chi_M \phi_\mu^\nu(f) = \phi_\mu^\nu(\chi_M f)$$

for every $\nu \succ \mu$. Thus $E(M)[f] = [\chi_M f]$ defines a mapping on $L_2[J]$. (The mapping $E(M)$ is known, of course, as the *direct limit* of the system of standard spectral projections (multiplication by χ_M) on the various spaces $L_2[\mu]$.) It is a triviality to verify that $E(M)$ is a projection on $L_2[J]$ and that the assignment $M \to E(M)$ defines a spectral measure on $L_2[J]$. We shall call this the *standard* spectral measure on $L_2[J]$.

Suppose now that μ is one of the measures belonging to J. Then there is a natural unitary mapping of $L_2[\mu]$ into $L_2[J]$, viz., the mapping $f \to [f]$.

In this sense we may say that $L_2[\mu]$ "is" a subspace of $L_2[J]$. Note that under this identification the standard spectral measure on $L_2[\mu]$ is identified with the restriction of the standard spectral measure on $L_2[J]$ to $L_2[\mu]$ regarded as a subspace. In the same vein we observe that if μ and ν are two measures in J, then $L_2[\mu]$ and $L_2[\nu]$ are orthogonal (when viewed as subspaces of $L_2[J]$) when and only when $\mu \perp \nu$, and likewise that $L_2[\nu]$ "contains" $L_2[\mu]$ when and only when $\mu \prec \nu$. (These are routine verifications and are left to the reader.) In short, the assignment $\mu \longrightarrow L_2[\mu]$ (regarded as a subspace of $L_2[J]$) carries J isomorphically into the subspace lattice of $L_2[J]$. In this context it is important not to overlook the following fact.

PROPOSITION 11.2. *Let J be a σ-ideal in $\mathfrak{M}$ and let $\{\mu_\alpha\}$ be a pairwise singular system of generators for J. Then the (external) direct sum*

$$\mathcal{H} = \sum_\alpha \oplus L_2[\mu_\alpha]$$

is isometrically isomorphic to $L_2[J]$ under the natural mapping

$$\{f_\alpha\} \xrightarrow{\Phi} \sum_\alpha [f_\alpha].$$

Moreover, Φ transforms the direct sum of the various standard spectral measures on the spaces $L_2[\mu_\alpha]$ onto the standard spectral measure on $L_2[J]$.

PROOF. Since the spaces $L_2[\mu_\alpha]$ are pairwise orthogonal when regarded as subspaces of $L_2[J]$, it is clear that Φ is a unitary mapping, and it is a routine matter to verify that Φ intertwines the standard spectral measures as specified. Hence everything comes down to showing that Φ is onto. Suppose that $[f]$ is an element of $L_2[J]$ that is orthogonal to every $L_2[\mu_\alpha]$ and let $\nu = \mu_{[f]}$ (with respect to the standard spectral measure on $L_2[J]$). A check of the definitions discloses that ν belongs to J—indeed, J itself is the σ-ideal of the standard spectral measure on $L_2[J]$—but that $L_2[\nu]$ is orthogonal to every one of the subspaces $L_2[\mu_\alpha]$. But then ν is singular with respect to all of the measures μ_α, whence it follows that $\nu = 0$ and, finally, that $f = 0$ as well. Q.E.D.

Let us call a subspace $\mathcal{W}$ of $\mathcal{H}$ a *generalized cycle* with respect to a spectral measure E if there is a unitary equivalence between the restriction of E to $\mathcal{W}$ and a standard spectral measure on some space $L_2[J]$. (This unitary equivalence is not uniquely determined, of course, without making some kind of uncanonical normalization, but it is easy to see that a unitary operator on $L_2[J]$ that commutes with the standard spectral measure must leave all of the subspaces $L_2[\mu]$, $\mu \in J$, invariant, so it is unique enough.) Moreover, let us broaden the notion of *stack* to include subspaces that are direct sums of *generalized* cycles, all with the same σ-ideal J (which we shall call the σ-ideal of the stack, of course). Then it is clear that Theorem 8.4 admits the following simpler paraphrase.

THEOREM 8.4″. *Let* E *be a spectral measure on a Hilbert space* $\mathcal{H}$. *Then there exist pairwise singular* σ*-ideals* $\mathcal{J}_c$, *and corresponding subspaces* S_c, *for all cardinal numbers* c, $1 \leqslant c \leqslant \dim \mathcal{H}$, *such that each* S_c *is a stack of multiplicity* c *having* σ*-ideal* $\mathcal{J}_c$, *and such that* $\mathcal{H}$ *is the direct sum of the* S_c*'s*:

$$\mathcal{H} = \sum_c \oplus S_c.$$

In this decomposition everything is uniquely determined by E.

BIBLIOGRAPHY

1. William Arveson, *Representations of C*-algebras* (to appear).

2. Arlen Brown, *The unitary equivalence of binormal operators*, Amer. J. Math **76** (1954), 414–434. MR **15**, 967.

3. D. Deckard, *Complete sets of unitary invariants for compact and trace class operators*, Acta Sci. Math. (Szeged) **28** (1967), 9–20.

4. J. Dixmier, *Les algèbres d'opérateurs dans l'espace Hilbertien* (*Algèbres de von Neumann*), Cahiers scientifiques, fasc. 25, Gauthier-Villars, Paris, 1957. MR **20** #1234.

5. P. R. Halmos, *Introduction to Hilbert space and the theory of spectral multiplicity*, Chelsea, New York, 1951. MR **13**, 563.

6. E. Hellinger, *Die Orthogonalinvarianten quadratischer Formen*, Inaugural-Dissertation, Göttingen, 1907.

7. R. V. Kadison, *Unitary invariants for representations of operator algebras*, Ann. of Math. (2) **66** (1957), 304–379. MR **19**, 665.

8. F. D. Murnaghan, *On the unitary invariants of a square matrix*, Ann. Acad. Brasil. Ci. 26 (1954), 1–7. MR **16**, 211.

9. E. Nelson, *Topics in dynamics* I: *Flows*, Mathematical Notes, Princeton, 1969.

10. Carl Pearcy, *A complete set of unitary invariants for* 3×3 *complex matrices*, Trans. Amer. Math. Soc. **104** (1962), 425–429. MR **26** #2451.

11. ———, *A complete set of unitary invariants for operators generating finite W*-algebras of type* I, Pacific J. Math. **12** (1962), 1405–1416. MR **26** #6816.

12. L. Schwartz, *Généralisation des espaces* L^p, Publ. Inst. Statist. Univ. Paris **6** (1957), 241–250. MR **20** #6034.

13. I. E. Segal, *Decompositions of operator algebras*. I, II, Mem. Amer. Math. Soc. No. 9 (1951). MR **13**, 472.

14. W. Specht, *Zur Theorie der Matrizen*. II, Jber. Deutsch. Math. Verein. **50** (1940), 19–23. MR **2**, 118.

15. F. Wecken, *Unitärinvarianten selbstadjungierter Operatoren*, Math. Ann. **116** (1939), 422–455.

INDIANA UNIVERSITY

Mathematical Surveys
Volume 13
1974

IV

CANONICAL MODELS

BY

R. G. DOUGLAS

AMS (MOS) subject classifications (1970). Primary 47A45, 47A20; Secondary 46L15, 46J15, 46J25, 47A10, 47A15, 47A60, 47C05.

Preparation of this paper was supported in part by a Sloan Foundation Fellowship and a grant from the National Science Foundation.

R. G. DOUGLAS

Introduction. Few classes of operators on an infinite-dimensional Hilbert space can be said to be fully understood. Even the most optimistic accounting would list only the classes of selfadjoint and unitary operators. Indeed for most operators it is not yet clear what the appropriate questions are. Despite this the number of mathematicians working in operator theory has continued to grow, and a brief review of the recent monographs which have appeared in this area attests to both its health and vigor. There are many approaches to the study of operator theory, and perhaps the best way of distinguishing among them is by considering the point of view adopted.

Our principal goal in this exposition is to discuss the canonical models approach to operator theory. Although we have begun such an exposition elsewhere **[20]** [1], our treatment here will be quite different from that and is suggested, in part, by certain possible generalizations which we will mention later.

In addition to presenting the canonical model of Sz.-Nagy and Foiaş for contraction operators, we shall discuss their theory of C_0 operators. This class consists of operators possessing a minimum function analogous to the minimum polynomial for operators on finite-dimensional spaces and indeed the theory of C_0 operators now subsumes most of the results from linear algebra concerning such operators.

The principal reference for this exposition is the book of Sz.-Nagy and Foiaş [73] and some of their more recent papers **[70]**, **[71]**, **[72]**, **[74]** and **[75]**. In addition, we cite the works of certain other authors but make no attempt at compiling an exhaustive bibliography. Nonetheless, the interested reader should find sufficient information for further reading on the topics we mention.

We shall assume throughout that the reader has an understanding of the basic results from measure theory, complex variables, functional analysis and operator theory. Moreover, to avoid duplication we shall make reference to certain

---

[1] We mention that the author has no plan at present to publish the successor paper promised in **[20]**.

topics discussed in the preceding articles in this volume by Brown and Sarason.

Lastly, although each section builds on the earlier ones, the level and difficulty varies considerable.

We now proceed to set forth some ground rules for terminology. We consider only *complex* Hilbert spaces and *operator* means bounded linear operator. A *subspace* is a closed linear manifold and for M a subspace of the Hilbert space H, we let P_M denote the (orthogonal) projection of H onto M. If S is a subset of H, then we denote by clos $[S]$ the closure of S. A subspace M of H is said to be an *invariant subspace* for the operator T on H if $TM \subset M$. If we also have $TM^{\perp} \subset M^{\perp}$, then M is said to be a *reducing subspace* for T. Lastly, a subspace M is said to be a *full subspace* for T if the smallest reducing subspace for T containing M is H; or equivalently, M is a full subspace for T if and only if $M^{\perp}$ contains no nontrivial reducing subspace for T. Observe that M is a full subspace for T if and only if it is a full subspace for T^*.

1. Canonical models and unitary dilations. One method of studying operators on a finite-dimensional Hilbert space involves the Jordan canonical form. The characteristic polynomial of some matrix representing the operator is computed and its roots are determined. If the roots are distinct, then the operator is similar to the operator which has a diagonal matrix with these roots as diagonal entries. If multiple roots occur, then the situation is more complicated, and there is a direct sum of Jordan blocks corresponding to each root. Thus a "canonical model" is obtained for each operator and two operators are similar if and only if their canonical models coincide.

Despite the relative simplicity of this canonical model, it is not a panacea for all problems in linear algebra. For example, the Jordan canonical form is of no help in computing the numerical range of an operator, that is, the subset $\{(Tf, f): f \in H, \|f\| = 1\}$ of the complex numbers $\mathbf{C}$, where T is an operator on H. The difficulty is that the Jordan canonical form is a similarity model while the numerical range is a unitary invariant.

A canonical model for an operator is a "natural" representation of the operator in terms of simpler operators and in a context in which more structure is present. Such a representation need not make all answers transparent, and indeed not all questions need be answerable in terms of it. Rather a choice must be made between the generality of the class covered and the completeness of the information which the model provides. For example, the spectral theorem states that every normal operator on a finite-dimensional Hilbert space is unitarily equivalent to the operator having a diagonal matrix with entries equal to the roots of the characteristic polynomial counted multiply. And, in this case, the numerical range is easily seen to be the convex hull of these roots.

A canonical model for normal operators on arbitrary Hilbert spaces is also provided by the spectral theorem and goes a long way toward resolving problems concerning normal operators on infinite-dimensional Hilbert spaces. Most attempts at providing canonical models for nonnormal operators have centered around generalizing the Jordan canonical form. Since the existence of such a model implies that the operator possesses many invariant subspaces, this approach has been successful only for those few classes, such as the class of compact operators, for which the existence of invariant subspaces has been proved (cf. **[33]**).

The canonical model which we shall be discussing is of quite a different nature. Rather than attempting to decompose a given operator into a sum of more elementary operators, one seeks to represent the given operator as "part" of a naturally associated operator of a better understood type on a larger space. In particular, in the canonical model of Sz.-Nagy and Foiaş, one seeks to represent a given contraction operator as the "compression" of an associated unitary operator.

If T is a contraction operator ($\|T\| \leqslant 1$) on the Hilbert space $\mathcal{H}$, then the operator W defined on the direct sum $\mathcal{K} = \mathcal{H} \oplus \mathcal{H}$ by the operator matrix $\begin{pmatrix} T & D_{T^*} \\ D_T & T^* \end{pmatrix}$ is unitary, where D_T is the nonnegative square root of $I - T^*T$. The only nonobvious facts needed to verify this statement are the identities $TD_T = D_{T^*}T$ and $T^*D_{T^*} = D_TT^*$ which are established as follows. Since the identity $TD_T^{2k} = D_{T^*}^{2k}T$ holds for all nonnegative integers k, we have $Tp(D_T^2) = p(D_{T^*}^2)T$ for each polynomial $p(x)$. Choosing a sequence of real polynomials $\{p_n(x)\}_{n=1}^{\infty}$ which converges uniformly on the interval $[0, 1]$ to the square root function, we obtain in the limit that $T(D_T^2)^{1/2} = (D_{T^*}^2)^{1/2}T$, which is the desired identity.

If we identify $\mathcal{H}$ with the first component in the direct sum $\mathcal{K} = \mathcal{H} \oplus \mathcal{H}$, then $T = P_{\mathcal{H}}W|\mathcal{H}$, and thus T is the compression of the unitary operator W to the subspace $\mathcal{H}$. Thus every contraction can be obtained as the compression of a unitary operator. This observation and construction is due to Halmos **[34]** who seems to have initiated the study of compressions and dilations of operators. The relation between the contraction and the associated unitary operator is very tenuous, however, especially since there is no uniqueness.

It was Sz.-Nagy who observed **[67]** that if we assume that a stronger relationship exists between T and W, then the associated unitary dilation would be far more useful. What he asked was that the identity $T^n = P\, W^n|\mathcal{H}$ hold for every positive integer n. Indeed the existence of such a *unitary dilation* yields von Neumann's result that the unit disk is a spectral set for the given contraction. In **[67]**, Sz.-Nagy proved that every contraction possesses a unitary dilation in the preceding sense and, moreover, that such a unitary dilation is uniquely determined if we assume that it is "minimal."

A few years later J. J. Schäffer **[62]** showed that a remarkable and ingenious variant of the Halmos construction using doubly infinite operator matrices yields the existence of a unitary dilation for a given contraction. If T is a contraction on H, K is the doubly infinite direct sum $\Sigma_{n=-\infty}^{\infty} \oplus H$, and W is the operator defined on H by the operator matrix

$$\begin{pmatrix} \ddots & & & & 0 \\ & I & D_T & -T^* & \\ & & \boxed{T} & D_{T^*} & \\ & & & I & \\ 0 & & & & \ddots \end{pmatrix}$$

where the square around T indicates that it is the (0, 0)-entry, then an elementary computation involving matrix multiplication and the same identities as before shows that W is unitary and that $T^n = P_H W^n | H$ for $n \geqslant 0$. This proves that every contraction possesses a unitary dilation. The unitary dilation constructed above is not necessarily *minimal*; that is, H is not necessarily a full subspace for W. (Consider for example the case when T is itself a unitary operator.) If one restricts W to the smallest reducing subspace for W which contains H, then one obtains a minimal unitary dilation for T.

An important property of the minimal unitary dilation is its uniqueness. Suppose W_1 and W_2 are unitary operators on the Hilbert spaces K_1 and K_2 containing H such that $T^n = P_H W_i^n | H$ and such that H is a full subspace for W_i for $i = 1, 2$. To establish uniqueness we want to construct an isometric isomorphism Φ from K_1 onto K_2 which is the identity on H and which satisfies $\Phi W_1 = W_2 \Phi$. By minimality, the set of vectors $E_i = \{\Sigma_{k=-N}^{N} W_i^k f_k$: N a positive integer and $f_{-N}, f_{-N+1}, \cdots, f_N \in H\}$ forms a dense linear manifold in K_i for $i = 1, 2$. Moreover, if Φ is to satisfy $\Phi W_1 = W_2 \Phi$ and $\Phi | H = I_H$, then we must have $\Phi(\Sigma_{k=-N}^{N} W_1^k f_k) = \Sigma_{k=-N}^{N} W_2^k f_k$. It suffices to check whether the preceding definition makes sense, and for this it is enough to show that Φ is an isometry. (To show that Φ is well defined requires showing that $\Sigma_{k=-N}^{N} W_1^k f_k = 0$ implies $\Sigma_{k=-N}^{N} W_2^k f_k = 0$.)

Computing, we have

$$\left\|\sum_{k=-N}^{N} W_2^k f_k\right\|^2 = \sum_{j,k=-N}^{N} (W_2^j f_j, W_2^k f_k)$$

$$= \sum_{j,k=-N;\, j\geq k}^{N} (W_2^{j-k} f_j, f_k) + \sum_{j,k=-N;\, j<k}^{N} (f_j, W_2^{k-j} f_k)$$

$$= \sum_{j,k=-N;\, j\geq k}^{N} (T^{j-k} f_j, f_k) + \sum_{j,k=-N;\, j<k}^{N} (f_j, T^{k-j} f_k)$$

$$= \sum_{j,k=-N}^{N} (W_1^j f_j, W_1^k f_k) = \left\|\sum_{k=-N}^{N} W_1^k f_k\right\|^2$$

and hence Φ is an isometry from E_1 onto E_2. Thus Φ extends to an isometric isomorphism from K_1 and K_2 having the required properties.

We state these results as

THEOREM 1. *If T is a contraction on H, then there exists a unitary operator W defined on a Hilbert space K containing H and satisfying $T^n = P\ W^n|H$ for $n \geqslant 0$. Moreover, if W is required to be minimal, then it is unique.*

Unitary dilations can be viewed in several different ways, each suggesting possible generalizations. Firstly, we can think of a unitary dilation as dilating the entire abelian semigroup $I, T, T^2, \cdots$, and hence it is natural to inquire whether unitary dilations exist for more general abelian semigroups of contractions. An example due to Parrott [58] shows that the answer is, in general, negative and the problem of deciding for which semigroups it is possible is unresolved.

Secondly, one can view a unitary dilation as providing a dilation for the entire algebra of polynomials in the contraction. It is in this context that the spectral inequality of von Neumann occurs.

COROLLARY 1.1. *If T is a contraction on H and $p(z)$ is a complex polynomial, then $\|p(T)\| \leqslant \sup\{|p(z)|: |z| \leqslant 1\}$.*

The proof consists of the observation that $p(T) = P_H p(W)|H$, and hence

$$\|p(T)\| \leqslant \|p(W)\| = \sup\{|p(w)|: w \in \sigma(W)\} \leqslant \sup\{|p(z)|: |z| \leqslant 1\}.$$

Thus if $\mathbf{A}$ denotes the closed subalgebra of the algebra $C(\mathbf{T})$ of continuous complex functions on the unit circle $\mathbf{T}$ that is generated by the polynomials, then the mapping $\pi(p) = p(T)$ can be extended to all of $\mathbf{A}$ by the corollary. Thus, the mapping π defines a contractive representation of the function algebra $\mathbf{A}$. Contractive representations of the larger algebra $C(\mathbf{T})$ depend only on the operator corresponding to the identity function z on $\mathbf{T}$ which can be shown to be unitary. The unitary dilation W provides a dilation of the representation π to the representation ρ of $C(\mathbf{T})$ defined by $\rho(\varphi) = \varphi(W)$ for φ in $C(\mathbf{T})$. Again the problem of deciding for which function algebras such a dilation is always possible is unsolved but the answer is known to be affirmative for Dirichlet algebras (**[5]**, **[30]**).

Our main objective is the study of a contraction using its associated unitary dilation and, more specifically, its canonical model. Before we can do this we need to know more about the structure of unitary operators and, in particular, that portion of the theory which relates to canonical models. We begin by showing how to decompose a contraction into a unitary part and a "completely nonunitary part."

A contraction T on the Hilbert space $\mathcal{H}$ is said to be *completely nonunitary* if the only subspace $\mathcal{M}$ of $\mathcal{H}$ for which the compression $P_{\mathcal{M}}T|\mathcal{M}$ is unitary is $\mathcal{M} = \{0\}$. (It is sufficient to assume this only for $\mathcal{M}$ a reducing subspace for T.)

A notational device which will be useful in what follows is letting $T^{(n)}$ denote T^n for $n \geqslant 0$ and T^{*-n} for $n < 0$.

PROPOSITION 1.2. *If T is a contraction on $\mathcal{H}$, then there exist reducing subspaces $\mathcal{H}_1$ and $\mathcal{H}_2$ for T such that $\mathcal{H} = \mathcal{H}_1 \oplus \mathcal{H}_2$, $T|\mathcal{H}_1$ is completely nonunitary, and $T|\mathcal{H}_2$ is unitary.*

PROOF. Since T is a contraction, we have $\|T^{(n)*}T^{(n)}\| \leqslant 1$ and hence $I - T^{(n)*}T^{(n)}$ is a nonnegative operator. If we set $\mathcal{H}_2 = \bigcup_{n=-\infty}^{\infty} \ker [I - T^{(n)*}T^{(n)}]$, then $\mathcal{H}_2$ is a closed subspace of $\mathcal{H}$ and for f in $\mathcal{H}_2$ we have $(I - T^{(n)*}T^{(n)})f = 0$, which implies that $\|f\|^2 = (f, f) = (T^{(n)*}T^{(n)}f, f) = \|T^{(n)}f\|^2$ for each integer n. Conversely, if f is a vector in $\mathcal{H}$ satifsying $\|T^{(n)}f\| = \|f\|$ for each integer n, then f is in $\mathcal{H}_2$. Since $T^*Tf = TT^*f = f$ for f in $\mathcal{H}_2$, it is clear that $\mathcal{H}_2$, is a reducing subspace for T and that $T|\mathcal{H}_2$ is a unitary operator.

It remains only to show that $T|\mathcal{H}_1$ is completely nonunitary, where $\mathcal{H}_1 = \mathcal{H} \ominus \mathcal{H}_2$. If $\mathcal{M}$ is a subspace of $\mathcal{H}$ such that $P_{\mathcal{M}}T|\mathcal{M}$ is unitary, then for f in $\mathcal{M}$ we have

$$\|f\|^2 = \|P_{\mathcal{M}}Tf\|^2 \leqslant \|P_{\mathcal{M}}Tf\|^2 + \|(I - P_{\mathcal{M}})Tf\|^2 = \|Tf\|^2 \leqslant \|f\|^2.$$

Hence, we see that $(I - P_{\mathcal{M}})Tf = 0$ for f in $\mathcal{M}$, and therefore $\mathcal{M}$ is an invariant subspace for T. Moreover, since $P_{\mathcal{M}}T^*|\mathcal{M} = (P_{\mathcal{M}}T|\mathcal{M})^*$ is also unitary, we see that $\mathcal{M}$ is a reducing subspace for T. Finally, we see that $\|T^n f\| = \|T^{*n} f\| = \|f\|$ for f in $\mathcal{M}$, and thus $\mathcal{M}$ is contained in $\mathcal{H}_2$, which completes the proof.

This decomposition results from the extremal character of the unitary property for contractions.

If W is a unitary operator on the Hilbert space $\mathcal{K}$, $\mathcal{H}$ a subspace of $\mathcal{K}$, and T the compression of W to $\mathcal{H}$, then it is highly unlikely that $T^2 = P_{\mathcal{H}} W^2 | \mathcal{H}$, let alone $T^n = P_{\mathcal{H}} W^n | \mathcal{H}$ for all positive integers. For the latter to hold $\mathcal{H}$ must be a very special subspace relative to W. If $\mathcal{H}$ is invariant for W, then $T = P_{\mathcal{H}} W | \mathcal{H} = W | \mathcal{H}$ and hence $T^n = W^n | \mathcal{H} = P_{\mathcal{H}} W^n | \mathcal{H}$. Dually, if $\mathcal{H}$ is invariant for W^*, then $T^n = (T^{*n})^* = (P_{\mathcal{H}} W^{*n} | \mathcal{H})^* = P_{\mathcal{H}} W^n | \mathcal{H}$. Moreover, it is possible to combine these two cases as follows. Suppose $\mathcal{M}$ and $\mathcal{N}$ are invariant subspaces for W, $\mathcal{M}$ is contained in $\mathcal{N}$, and $\mathcal{H} = \mathcal{N} \ominus \mathcal{M}$. If we write $\mathcal{K} = \mathcal{M}^\perp \oplus \mathcal{H} \oplus \mathcal{N}$, then the matrix for W relative to this decomposition is lower triangular, and from the usual definition of matrix multiplication it follows that $T^n = P_{\mathcal{H}} W^n | \mathcal{H}$.

A subspace which can be written as the difference $\mathcal{N} \ominus \mathcal{M}$ of two invariant subspaces $\mathcal{M}$ and $\mathcal{N}$ for W is said to be a *semi-invariant subspace* for W. Our preceding calculation shows that the compression of W to a semi-invariant subspace is a "power dilation." Sarason made this observation in **[59]** and established the nonobvious converse. The following proof of this fact is due to Foiaş (cf. **[59]**).

PROPOSITION 1.3. *If A is an operator on the Hilbert space $\mathcal{K}$, $\mathcal{H}$ is a subspace of $\mathcal{K}$, and $B = P_{\mathcal{H}} A | \mathcal{H}$, then $B^n = P_{\mathcal{H}} A^n | \mathcal{H}$ for each nonnegative integer n if and only if $\mathcal{H}$ is a semi-invariant subspace for A.*

PROOF. We just showed that $\mathcal{H}$ semi-invariant for A implies that $B^n = P_{\mathcal{H}} A^n | \mathcal{H}$ for $n \geqslant 0$.

Suppose we assume $B^n = P_{\mathcal{H}} A^n | \mathcal{H}$ for all nonnegative integers n. Let $\mathcal{N}$ be the smallest invariant subspace for A containing $\mathcal{H}$ and set $\mathcal{M} = \mathcal{N} \ominus \mathcal{H}$. To complete the proof we need only to show that $\mathcal{M}$ is an invariant subspace for A and for this it is sufficient to prove that $P_{\mathcal{M}} A P_{\mathcal{M}} = A P_{\mathcal{M}}$. Since $P_{\mathcal{M}} = P_{\mathcal{N}} - P_{\mathcal{H}}$ and the identities $P_{\mathcal{N}} A P_{\mathcal{N}} = A P_{\mathcal{N}}$ and $P_{\mathcal{N}} A P_{\mathcal{H}} = A P_{\mathcal{H}}$ are obvious, this is equivalent to showing that $P_{\mathcal{H}} A P_{\mathcal{H}} = P_{\mathcal{H}} A P_{\mathcal{N}}$. If $p(z)$ is a polynomial and f is a vector in $\mathcal{H}$, then

$$P_{\mathcal{H}} A P_{\mathcal{H}} p(A) f = P_{\mathcal{H}} A P_{\mathcal{H}} p(A) P_{\mathcal{H}} f = P_{\mathcal{H}} A p(A) f,$$

where we have used the fact that A is a power dilation of B. Since the closed linear span of the vectors of the form $p(A)f$ is N, it follows that $P_H A P_H = P_H A P_H P_N = P_H A P_N$ which completes the proof.

There is an equivalent form of this proposition which we shall find useful.

COROLLARY 1.4. *If A is an operator on the Hilbert space K, H is a subspace of K, and $B = P_H A | H$, then $B^n = P_H A^n | H$ for each nonnegative integer n if and only if there exist orthogonal subspaces G and G_* which are invariant for A and A^*, respectively, and such that $K = G_* \oplus H \oplus G$.*

PROOF. In the proposition set $G = M$ and $G_* = N^\perp$.

Using this result we see that the study of a contraction T with unitary dilation W can be reduced to the study of the invariant subspaces of W and W^*. In order to continue we must consider the invariant subspaces of unitary operators. In the next section we describe the representation of unitary operators as "multiplication operators" on a direct integral space.

2. **Unitary operators and invariant subspaces.** If μ is a finite nonnegative Borel measure on $\mathbf{T}$, then the multiplication operator M_z defined on $L^2(\mu)$ by $M_z f = zf$ is unitary and one form of the spectral theorem yields (cf. [23], [61]) the fact that every unitary operator possessing a cyclic vector is unitarily equivalent to an operator of this form. The unitary operator $M_z \oplus M_z$ defined on $L^2(\mu) \oplus L^2(\mu)$ does not possess a cyclic vector and hence we must consider more general models to encompass all unitary operators. If we consider operators of the form $\Sigma_n \oplus M_z$ acting on the countable direct sum $\Sigma_n \oplus L^2(\mu_n)$, then we obtain a model for the most general unitary operator acting on a separable Hilbert space.

This model is not, however, suitable for our purposes. The difficulty is that we have chosen a basis for the eigenspaces and more importantly, one which is irrelevant to our interests here. To avoid this we must introduce the notion of a direct integral due to von Neumann. Our approach is somewhat different from his, and is based on [21]. We omit most of the rather technical details and refer the reader to the slightly different presentation of these results in [6].

We begin by briefly recalling the notion of a direct sum and its relation to "diagonalizable" normal operators to motivate our definition of direct integral. Let Λ be a bounded countable subset of the complex numbers $\mathbf{C}$ and let E_λ be a separable Hilbert space for each λ in Λ. The product space

$$\prod_{\lambda \in \Lambda} E_\lambda = \left\{ f\colon \Lambda \to \bigcup_{\lambda \in \Lambda} E_\lambda,\ f(\lambda) \in E_\lambda \right\}$$

is a linear space under pointwise addition, and we obtain the direct sum $\Sigma_{\lambda\in\Lambda} \oplus E_\lambda$ by considering those functions f in $\Pi_{\lambda\in\Lambda} E_\lambda$ satisfying the growth condition $\Sigma_{\lambda\in\Lambda} \|f(\lambda)\|^2 < \infty$. Further, the multiplication operator M_z defined by $(M_z f)(\lambda) = \lambda f(\lambda)$ for f in $\Sigma_{\lambda\in\Lambda} \oplus E_\lambda$ is normal and E_λ can be identified with the eigenspace for M_z corresponding to the eigenvalue λ. Moreover, the operators that commute with M_z have an extremely useful representation on $\Sigma_{\lambda\in\Lambda} \oplus E_\lambda$; namely, if T commutes with M_z, then there exists an operator T_λ defined on E_λ for λ in Λ, such that $T = \Sigma_{\lambda\in\Lambda} \oplus T_\lambda$. In particular, a subspace M reduces M_z if and only if there exists a subspace M_λ for each E_λ, respectively, such that $M = \{f \in \Sigma_{\lambda\in\Lambda} \oplus E_\lambda \colon f(\lambda) \in M_\lambda\}$. (The reader should attempt to describe the reducing subspaces for a diagonal normal operator acting on the direct sum of one-dimensional spaces; this latter representation corresponds to making a choice of a basis for the eigenspaces.)

A direct integral can be viewed as a "continuous direct sum." Assume that Λ is a compact subset of $\mathbf{C}$ and that ν is a positive Borel measure with closed support equal to Λ. (In the previous paragraph μ would be the counting measure on Λ with closed support equal to the closure of Λ.) Again assume that E_λ is a separable Hilbert space for each λ in Λ and that the real function $\dim E_\lambda$ is ν-measurable and nonzero ν-a.e. The heuristic idea is to consider the collection of square-integrable ν-measurable functions in $\Pi_{\lambda\in\Lambda} E_\lambda$. Since the notion of a function in $\Pi_{\lambda\in\Lambda} E_\lambda$ being ν-measurable does not make sense, we must proceed differently.

If we set $\mathbf{N} = \{f \in \Pi_{\lambda\in\Lambda} E_\lambda \colon f(\lambda) = 0 \ \nu\text{-a.e.}\}$, then $\mathbf{N}$ is a subspace of $\Pi_{\lambda\in\Lambda} E_\lambda$, and the quotient $\mathbf{F} = \Pi_{\lambda\in\Lambda} E_\lambda/\mathbf{N}$ is a linear space called the field of Hilbert spaces associated with $(\Lambda, \nu, \{E_\lambda\}_{\lambda\in\Lambda})$. As usual we speak of the elements of $\mathbf{F}$ as functions. We let $\mathbf{F}_2$ denote the subset of functions f in $\mathbf{F}$ for which the function $\|f(\lambda)\|_{E_\lambda}$ is both ν-measurable and square-integrable. Since it is possible for the functions $\|f(\lambda)\|^2_{E_\lambda}$ and $\|g(\lambda)\|^2_{E_\lambda}$ to be ν-measurable while the function $\|(f+g)(\lambda)\|^2_{E_\lambda}$ is not, the set $\mathbf{F}_2$ is not a linear subspace of $\mathbf{F}$ unless ν is atomic.

We are interested in linear submanifolds of $\mathbf{F}_2$ and, in particular, in maximal ones with respect to inclusion. If $\mathbf{M}$ is a maximal linear submanifold, then the polarization formula for the inner product shows that the function $(f(\lambda), g(\lambda))_{E_\lambda}$ is ν-integrable for f and g in $\mathbf{M}$ and hence we can define the inner product $(f, g) = \int_\Lambda (f(\lambda), g(\lambda))_{E_\lambda}\, d\nu$ on $\mathbf{M}$. Moreover using the maximality of $\mathbf{M}$ we can prove that $\mathbf{M}$ is complete with respect to this inner product and hence a Hilbert space. Further, $\mathbf{M}$ is an $L^\infty(\nu)$-module; that is, φf is in $\mathbf{M}$ for φ in $L^\infty(\nu)$ and f in $\mathbf{M}$, where $(\varphi f)(\lambda) = \varphi(\lambda) f(\lambda)$.

Now $\mathbf{M}$ is not unique unless ν is atomic and $\mathbf{M} = \mathbf{F}_2$. The nonuniqueness, however, is not serious if we limit our attention to those maximal linear submanifolds which define separable Hilbert spaces, *which we do*! (Unfortunately, nonseparable examples exist.) If V_λ is a unitary operator on E_λ for λ in Λ, then we can define V on $\mathbf{F}_2$ by $(Vf)(\lambda) = V_\lambda f(\lambda)$ for f in $\mathbf{F}_2$. Since V obviously takes linear submanifolds onto linear submanifolds, it follows that $V\mathbf{M}$ is a separable maximal linear submanifold if $\mathbf{M}$ is. And what is more important, the converse is valid. Thus the separable maximal linear submanifold is unique up to such an operator V. We let $\int_\Lambda \oplus E_\lambda \, d\nu$ denote any fixed maximal linear submanifold associated with $(\Lambda, \nu, \{E_\lambda\}_{\lambda\in\Lambda})$. This completes the basic construction of the direct integral. We now proceed to set forth some of its important properties.

There is a natural collection of "multiplication operators" that can be defined on the direct integral space. For φ a function in $L^\infty(\nu)$, define the multiplication operator M_φ on $\int_\Lambda \oplus E_\lambda \, d\nu$ by $(M_\varphi f)(\lambda) = \varphi(\lambda) f(\lambda)$ for f in $\int_\Lambda \oplus E_\lambda \, d\nu$. Moreover, the mapping $\varphi \longrightarrow M_\varphi$ can be shown to be a $*$-isomorphism from $L^\infty(\nu)$ onto the algebra $\mathfrak{A} = \{M_\varphi : \varphi \in L^\infty(\nu)\}$ of operators acting on $\int_\Lambda \oplus E_\lambda \, d\nu$. The operator M_z can be viewed as the basic operator action on the direct integral space. In particular, $\mathfrak{A}$ is the smallest weakly closed $*$-algebra of operators on $\int_\Lambda \oplus E_\lambda \, d\nu$ which contains M_z and can be identified using the Fuglede theorem and the von Neumann double commutant theorem with the algebra of all operators commuting with every operator that commutes with M_z (cf. **[6]**).

Let us now consider how the direct integral depends on the various components used to define it. If $\{E'_\lambda\}_{\lambda\in\Lambda}$ is another assignment of a separable Hilbert space E'_λ to each λ in Λ such that the function $\dim E'_\lambda$ is ν-measurable and nonzero ν-a.e., then we have the direct integral $\int_\Lambda \oplus E'_\lambda \, d\nu$ defined. Moreover, if $\dim E_\lambda = \dim E'_\lambda$ ν-a.e., then there exists an isometric isomorphism Ψ_λ from E_λ onto E'_λ ν-a.e., and we can define Ψ from $\Pi_{\lambda\in\Lambda} E_\lambda$ to $\Pi_{\lambda\in\Lambda} E'_\lambda$ pointwise. Moreover, it is clear that Ψ takes maximal linear submanifolds of $\mathbf{F}_2$ onto maximal linear submanifolds of $\mathbf{F}'_2$ and hence we can choose Ψ to map $\int_\Lambda \oplus E_\lambda \, d\nu$ onto $\int_\Lambda \oplus E'_\lambda \, d\nu$. Further, Ψ is an isometric isomorphism and $M_\varphi \Psi = \Psi M_\varphi$ for φ in $L^\infty(\nu)$. Thus the two direct integrals are entirely equivalent. Moreover, if ν_1 is a measure on Λ in the same measure class as ν, that is, ν and ν_1 are mutually absolutely continuous, then the direct integral $\int_\Lambda \oplus E_\lambda \, d\nu_1$ can also be formed. However, the multiplication operator Φ defined $(\Phi f)(\lambda) = (d\nu_1/d\nu)^{1/2}(\lambda) f(\lambda)$ establishes an isometric isomorphism between $\int_\Lambda \oplus E_\lambda \, d\nu$ and $\int_\Lambda \oplus E_\lambda \, d\nu_1$. Further, Φ satisfies the relation

$M_\varphi \Phi = \Phi M_\varphi$ for every φ in $L^\infty(\nu) = L^\infty(\nu_1)$. Thus the important parameters in defining a direct integral are Λ, the measure class of ν, and the dimension function $\dim \mathcal{E}_\lambda$. The set Λ can be identified with the spectrum of M_z, a measure ν in the appropriate measure class is called a *scalar spectral measure* for M_z, and $\dim \mathcal{E}_\lambda$ is the *multiplicity function* for M_z. Lastly, two normal operators are unitarily equivalent if and only if their scalar spectral measures are mutually absolutely continuous and their multiplicity functions are equal almost everywhere with respect to the scalar spectral measure. (See, in this connection, the article by Arlen Brown in this volume.)

If the dimension function is constant ν-a.e., then an equivalent direct integral can be obtained by taking each $\mathcal{E}_\lambda$ equal to a fixed Hilbert space $\mathcal{E}$. In this case the field $\mathbf{F}$ can be identified as all functions from Λ to $\mathcal{E}$ and $\mathbf{F}_2$ has a similar interpretation. Moreover, although any maximal linear submanifold can be used to define the direct integral $\int_\Lambda \bigoplus \mathcal{E}_\lambda \, d\nu$, there is now a natural candidate: The collection of functions f in $\mathbf{F}_2$ for which the function $(f(\lambda), e)$ is ν-measurable for each vector e in $\mathcal{E}$. We denote this direct integral by $L^2_{\mathcal{E}}(\nu)$.

The principal importance of the direct integral decomposition for normal operators, beyond providing a representation for every normal operator along with a complete set of unitary invariants, is that the commutant of the operator M_z can be conveniently represented on it. Let $\int_\Lambda \bigoplus L(\mathcal{E}_\lambda) d\nu$ denote the collection of operator functions Φ from Λ to $\bigcup_{\lambda \in \Lambda} L(\mathcal{E}_\lambda)$ such that $\Phi(\lambda)$ lies in $L(\mathcal{E}_\lambda)$ for λ in Λ, the function $\|\Phi(\lambda)\|_{L(\mathcal{E}_\lambda)}$ is essentially bounded, and the function defined $(\Phi f)(\lambda) = \Phi(\lambda) f(\lambda)$ lies in $\int_\Lambda \bigoplus \mathcal{E}_\lambda \, d\nu$ for each f in $\int_\Lambda \bigoplus \mathcal{E}_\lambda \, d\nu$. (Observe that changing maximal linear submanifolds from $\mathbf{M}$ to $V\mathbf{M}$ causes the set $\int_\Lambda \bigoplus L(\mathcal{E}_\lambda) d\nu$ to be replaced by $V \int_\Lambda \bigoplus L(\mathcal{E}_\lambda) d\nu V^*$, and hence again the nonuniqueness if not serious.) A function Φ in $\int_\Lambda \bigoplus L(\mathcal{E}_\lambda) d\nu$ defines a bounded operator M_Φ defined on $\int_\Lambda \bigoplus \mathcal{E}_\lambda \, d\nu$ in the obvious way and this collection of operators forms the commutant of M_z (cf. **[6]** for the proof of this).

We began this discussion of direct integrals to study the invariant subspaces of unitary operators. If we specialize the preceding discussion to unitary operators, then Λ can be replaced by a closed subset of $\mathbf{T}$. Hence the desired representation for a unitary operator W on a Hilbert space $\mathcal{K}$ is a multiplication operator M_z acting on the direct integral $\int_\Lambda \bigoplus \mathcal{E}_\lambda \, d\nu$. We summarize our discussion for unitary operators in the following proposition. An analogous statement is valid for normal operators.

PROPOSITION 2.1. *If W is a unitary operator on the separable Hilbert* *space* $\mathcal{K}$, *there exists a positive Borel measure* ν *on a closed subset* Λ *of* $\mathbf{T}$ *and a family* $\{\mathcal{E}_\lambda\}_{\lambda\in\Lambda}$ *of separable Hilbert spaces such that the function* $\dim \mathcal{E}_\lambda$ *is* ν*-measurable and nonzero* ν*-a.e. and such that* W *is unitarily equivalent to* M_z *acting on* $\int_\Lambda \oplus \mathcal{E}_\lambda \, d\nu$. *Moreover, the measure class of* ν *and the function* $\dim \mathcal{E}_\lambda$ *form a complete set of unitary invariants for* M_z. *Lastly, the commutant of* M_z *is the algebra of operators* $\int_\Lambda \oplus L(\mathcal{E}_\lambda)\, d\nu$.

Now we proceed to describe the invariant subspaces of M_z. Usually the simplest invariant subspaces for an operator are the reducing subspaces and these may be viewed as the ranges of the (orthogonal) projections which commute with the operator. Since an arbitrary operator in the commutant of M_z has the form M_Φ, it is sufficient to decide which of these operators are projections. Observing that $M_\Phi^2 = M_{\Phi^2}$ and $M_\Phi^* = M_{\Phi^*}$, we see that M_Φ is a projection if and only if $\Phi(\lambda)$ is a projection ν-a.e. If we let $\mathcal{R}_\lambda$ denote the range of $\Phi(\lambda)$ in $\mathcal{E}_\lambda$, then the range of Φ can be identified with $\int_\Lambda \oplus \mathcal{R}_\lambda\, d\nu$, that is, the functions f in $\int_\Lambda \oplus \mathcal{E}_\lambda\, d\nu$ for which $f(\lambda)$ lies in $\mathcal{R}_\lambda$ ν-a.e. Conversely, if we let $\mathcal{Q}_\lambda$ denote a subspace of $\mathcal{E}_\lambda$ for each λ in Λ, then the collection of f in $\int_\Lambda \oplus \mathcal{E}_\lambda\, d\nu$ for which $f(\lambda)$ lies in $\mathcal{Q}_\lambda$ ν-a.e. forms a reducing subspace $\mathcal{Q}$ for M_z. It is, however, not necessarily the case that $\mathcal{Q} = \int_\Lambda \oplus \mathcal{Q}_\lambda\, d\nu$ since the function defined by $\Phi(\lambda) = P_{\mathcal{Q}_\lambda}$ need not be in $\int_\Lambda \oplus L(\mathcal{E}_\lambda)\, d\nu$. Thus it makes sense to restrict one's attention to the *range functions* $\mathcal{Q}_\lambda$ for which the operator function $P_{\mathcal{Q}_\lambda}$ lies in $\int_\Lambda \oplus L(\mathcal{E}_\lambda)\, d\nu$. And if we identify range functions which coincide ν-a.e., then one obtains the following result.

PROPOSITION 2.2. *Every reducing subspace of the operator* M_z *on* $\int_\Lambda \oplus \mathcal{E}_\lambda\, d\nu$ *is of the form* $\int_\Lambda \oplus \mathcal{R}_\lambda\, d\nu$ *for some range function* $\mathcal{R}_\lambda$. *Moreover, two range functions* $\mathcal{R}_\lambda$ *and* $\mathcal{R}'_\lambda$ *define the same reducing subspace if and only if* $\mathcal{R}_\lambda = \mathcal{R}'_\lambda$ ν*-a.e.*

If $\mathcal{G}$ is an invariant subspace for the unitary operator M_z on $\int_\Lambda \oplus \mathcal{E}_\lambda\, d\nu$, then there exist invariant subspaces $\mathcal{G}'$ and $\mathcal{G}''$ for M_z *such that* $\mathcal{G} = \mathcal{G}' \oplus \mathcal{G}''$, $\mathcal{G}'$ is reducing, and $\mathcal{G}''$ is *pure*; that is, $\mathcal{G}''$ contains no nonzero reducing subspace for M_z. For unitary operators this decomposition is often referred to as the Wold decomposition although it is due originally to von Neumann. The existence of such a decomposition is established for subnormal (and hence isometric) operators in the earlier article by Sarason **[61]**.

Since the reducing subspace $\mathcal{G}'$ is of the form $\int_\Lambda \oplus \mathcal{G}'_\lambda\, d\nu$ for some range function $\mathcal{G}'_\lambda$, we see that $\mathcal{G}''$ is contained in $(\int_\Lambda \oplus \mathcal{G}'_\lambda\, d\nu)^\perp = \int_\Lambda \oplus (\mathcal{G}'_\lambda)^\perp\, d\nu$. Therefore, the two pieces of the invariant subspace are contained in separate parts of the direct integral

$$\int_\Lambda \oplus E_\lambda \, d\nu = \int_\Lambda \oplus G'_\lambda \, d\nu \oplus \int_\Lambda \oplus (G'_\lambda)^\perp \, d\nu.$$

Thus the study of general invariant subspaces for M_z can be reduced to the pure case.

How does one describe the pure invariant subspaces for a unitary operator? In [**35**] Halmos showed (cf. [**61**]) that a pure invariant subspace M for W is determined by the subspace $\mathcal{D} = M \ominus WM$. In particular, we obtain that the subspaces $\{W^k \mathcal{D}\}_{k=0}^\infty$ are pairwise orthogonal (and hence $\mathcal{D}$ is said to be *wandering* for W) and that $M = \Sigma_{k=0}^\infty \oplus W^k \mathcal{D}$. Thus the problem of characterizing the pure invariant subspaces for M_z on $\int_\Lambda \oplus E_\lambda \, d\nu$ can be reduced to the determination of the wandering subspaces for M_z in terms of the direct integral representation.

Assume that $\mathcal{D}$ is a wandering subspace for M_z acting on $\int_\Lambda \oplus E_\lambda \, d\nu$ and that $d\nu = h d\lambda + d\sigma$ is the Lebesgue decomposition of ν; that is, σ is a singular measure on $\mathbf{T}$ and h is a nonnegative ν-measurable function which vanishes σ-a.e. If f is a function in $\mathcal{D}$, then

$$\int_\Lambda e^{in\lambda} (f(\lambda), f(\lambda))_{E_\lambda} \, d\nu(\lambda) = (M_z^n f, f) = 0 \quad \text{for } n > 0.$$

Since the measure ξ defined by $d\xi = \|f(\lambda)\|_{E_\lambda}^2 \, d\nu$ is positive, it follows that only the zeroth Fourier-Stieltjes coefficient can be different from zero and hence

$$c\, d\lambda = d\xi = \|f(\lambda)\|_{E_\lambda}^2 (h d\lambda + d\sigma)$$

for some positive constant c. Therefore, we see that $f(\lambda) = 0$ σ-a.e. and $\|f(\lambda)\|_{E_\lambda} = h^{-1/2}(\lambda)$ ν-a.e. In particular, the existence of a nontrivial pure invariant subspace for M_z implies that Lebesgue measure is absolutely continuous with respect to ν and hence $\Lambda = \mathbf{T}$. Since we may change measures so long as the measure class stays the same, we assume from now on that $d\nu = d\lambda/2\pi + d\sigma$. (This is possible since h vanishing on a set of positive Lebesgue measure implies that M_z has no pure invariant subspaces.) Now let C be a Hilbert space and $\Psi(\lambda)$ be an isometry from C into E_λ for each λ in $\mathbf{T}$ such that the function $\Psi(\lambda)b$ lies in $\int_\mathbf{T} \oplus E_\lambda \, d\nu$ for each vector b in C and $\Psi(\lambda) = 0$ σ-a.e. (Such a function will be called a *unitary function.*) Then the collection $\Psi C = \{\Psi(\lambda)b \colon b \in C\}$ is a wandering subspace for M_z as a simple computation shows. More importantly, every wandering subspace is of this form.

PROPOSITION 2.3. *A subspace $\mathcal{D}$ of $\int_\mathbf{T} \oplus E_\lambda \, d\nu$ is wandering for M_z if and only if there exists a Hilbert space C and a unitary function Ψ such that each $\Psi(\lambda)$ is an isometry from C into E_λ and such that ΨC lies in $\int_\mathbf{T} \oplus E_\lambda \, d\nu$ and equals $\mathcal{D}$.*

PROOF. Assume that $\mathcal{D}$ is a wandering subspace for M_z and that $\{d_k\}_{k=1}^N$ is an orthonormal basis for $\mathcal{D}$, where $1 \leqslant N \leqslant \infty$. Since $d_k(\lambda) = 0$ σ-a.e. by our previous remarks and

$$\int_{\mathbf{T}} e^{in\lambda}(d_k(\lambda), d_j(\lambda))\frac{d\lambda}{2\pi} = (M_z^n d_k, d_j) = \begin{cases} 1, & \text{if } n = 0,\ k = j, \\ 0, & \text{otherwise}, \end{cases}$$

we see that

$$(d_k(\lambda), d_j(\lambda)) = \begin{cases} 1\ d\theta\text{-a.e.} & \text{if } k = j, \\ 0\ d\theta\text{-a.e.} & \text{if } k \neq j. \end{cases}$$

Thus there exists a set E of Lebesgue measure zero such that $\{d_k(\lambda)\}_{k=1}^N$ is an orthonormal subset of $\mathcal{E}_\lambda$ for λ not in E. Hence, if $\mathcal{C}$ is a Hilbert space of the same dimension as $\mathcal{D}$ having the orthonormal basis $\{b_k\}_{k=1}^N$, then we can define the isometry $\Psi(\lambda)$ from $\mathcal{C}$ into $\mathcal{E}_\lambda$ such that $\Psi(\lambda)b_k = d_k(\lambda)$ for λ not in E. Defining $\Psi(\lambda)$ arbitrarily on E, we obtain Ψ of the prescribed form such that $\mathcal{D} = \Psi\mathcal{C}$.

COROLLARY 2.4. *If $\mathcal{C}$ and $\mathcal{C}'$ are Hilbert spaces and Ψ and Ψ' are unitary functions from $\mathcal{C}$ and $\mathcal{C}'$, respectively, to $\int_{\mathbf{T}} \oplus \mathcal{E}_\lambda\, d\nu$, then $\Psi\mathcal{C} = \Psi'\mathcal{C}'$ if and only if there exists an isometrical isomorphism C from $\mathcal{C}$ onto $\mathcal{C}'$ such that $\Psi = \Psi' C$.*

PROOF. Since the one direction is clear, assume that $\Psi\mathcal{C} = \Psi'\mathcal{C}'$. Then for each b in $\mathcal{C}$ there exists b' in $\mathcal{C}'$ such that $\Psi(\lambda)b = \Psi'(\lambda)b'$ ν-a.e. and defining $Cb = b'$ gives the required map.

There is a special case of the latter deserving individual attention. If $d\nu = d\lambda/2\pi$ and the dimension of $\mathcal{E}_\lambda$ is constant, then the direct integral in this case is the usual vector-valued Lebesgue space $\mathbf{L}^2_{\mathcal{E}}$ defined on the circle. The unitary operator $U = M_z$ defined on $\mathbf{L}^2_{\mathcal{E}}$ is called a *bilateral shift* and is characterized by the fact that it has Lebesgue measure as a scalar spectral measure and a constant multiplicity function. Moreover, the collection of constant functions in $\mathbf{L}^2_{\mathcal{E}}$ forms a natural wandering subspace, also denoted $\mathcal{E}$, for U which is *complete*, that is which satisfies $\mathbf{L}^2_{\mathcal{E}} = \Sigma_{n=-\infty}^{\infty} \oplus U^n\mathcal{E}$. It is the shift action of U relative to this decomposition which leads to the name "bilateral shift." Observe that U has many complete wandering subspaces and hence, strictly speaking, labeling a unitary operator a bilateral shift requires singling out a particular one.

The invariant subspace for U associated with the wandering subspace

$\mathcal{E}$ is $\mathbf{H}^2_{\mathcal{E}} = \Sigma_{n=0}^{\infty} \oplus U^n \mathcal{E}$. An alternate definition is that $\mathbf{H}^2_{\mathcal{E}}$ consists of those functions in $\mathbf{L}^2_{\mathcal{E}}$ having vanishing negative Fourier coefficients, that is, for which

$$\int_{\mathbf{T}} e^{in\lambda} f(\lambda) \frac{d\lambda}{2\pi} = 0 \quad \text{for } n > 0.$$

Thus $\mathbf{H}^2_{\mathcal{E}}$ is a vector-valued Hardy space and hence can also be identified with a space of vector-valued analytic functions on D having a norm square-summable Taylor series (cf. [38] and [73]). The operator U_+ obtained by restricting U to $\mathbf{H}^2_{\mathcal{E}}$ is called the *unilateral shift* and is an isometry. Using the aforementioned Wold decomposition, it is possible to show (cf. [61]) that every isometry is the direct sum of such a unilateral shift and a unitary operator.

It is also convenient to have notation for $\mathbf{L}^2_{\mathcal{E}} \ominus \mathbf{H}^2_{\mathcal{E}}$ and thus we set $\mathbf{K}^2_{\mathcal{E}} = \mathbf{L}^2_{\mathcal{E}} \ominus \mathbf{H}^2_{\mathcal{E}}$. This subspace is a pure invariant subspace for U^* and can be identified as a space of holomorphic functions on the exterior of $\mathbf{D}$ which vanish at infinity.

One reason for the importance of the Hardy spaces is that they are the "universal" pure invariant subspaces for unitary operators. The following result expresses this fact and is a mild generalization of the results of Beurling [7], Lax [47], [48] Halmos [35], and Helson and Lowdenslager [40].

THEOREM 2. *A subspace $\mathcal{G}$ of $\int_{\mathbf{T}} \oplus \mathcal{E}_\lambda \, dv$ is a pure invariant subspace for M_z if and only if there is an isometry Θ from some Hardy space $\mathbf{H}^2_{\mathcal{C}}$ onto $\mathcal{G}$ such that $\Theta U_+ = M_z \Theta$. Moreover, such a mapping Θ is called a unitary function, and there exists an isometry $\Theta(\lambda)$ from $\mathcal{C}$ into $\mathcal{E}_\lambda$ for λ in $\mathbf{T}$ such that $(\Theta f)(\lambda) = \Theta(\lambda) f(\lambda)$ for f in $\mathbf{H}^2_{\mathcal{C}}$.*

PROOF. If $\Theta(\lambda)$ is an isometry from $\mathcal{C}$ into $\mathcal{E}_\lambda$ for each λ in $\mathbf{T}$, then defining $(\Theta f)(\lambda) = \Theta(\lambda) f(\lambda)$ for f in $\mathbf{H}^2_{\mathcal{C}}$ yields an isometry from $\mathbf{H}^2_{\mathcal{C}}$ into $\int_{\mathbf{T}} \oplus \mathcal{E}_\lambda \, dv$. Further, the identity

$$(\Theta U_+ f)(\lambda) = \Theta(\lambda)\lambda f(\lambda) = \lambda(\Theta f)(\lambda) = (M_z \Theta f)(\lambda)$$

shows that $\Theta U_+ = M_z \Theta$ and thus the range $\mathcal{G}$ of Θ is an invariant subspace for M_z. Finally, the fact that $M_z^n \mathcal{G} = \Theta[U_+^n \mathbf{H}^2_{\mathcal{C}}]$ implies that $\bigcap_{n=0}^{\infty} M_z^n \mathcal{G} = \{0\}$ and therefore that $\mathcal{G}$ is a pure invariant subspace for M_z.

Conversely, assume that $\mathcal{G}$ is a pure invariant subspace for M_z with associated wandering subspace $\mathcal{D}$. If Ψ is a unitary function and $\mathcal{C}$ is a Hilbert space for which $\mathcal{D} = \Psi \mathcal{C}$, then it follows that $(\Theta f)(\lambda) = \Psi(\lambda) f(\lambda)$ defines a unitary function mapping $\mathbf{H}^2_{\mathcal{C}}$ onto $\mathcal{G}$ and possessing the desired properties.

If the direct integral is $\mathbf{L}^2_E$, then unitary functions have a simpler representation. Since the dimension of C is not greater than the dimension of E, we can identify C with a subspace of E. If we extend $\Theta(\lambda)$ to be 0 on $E \ominus C$, then we obtain a function Θ' from E to E for which each $\Theta'(\lambda)$ is a partial isometry. If E is one dimensional, then Θ' is a scalar function and we recover the result of Helson and Lowdenslager (cf. **[40]**). Further, if G happens to be contained in $\mathbf{H}^2_E$, then Θ' has a holomorphic extension into **D**. Lastly, observe that the associated wandering subspace is complete if and only if Θ' is unitary-valued. Such an invariant subspace is said to have *full range* **[38]**. If an invariant subspace is contained in $\mathbf{H}^2_E$ and has full range, then the associated holomorphic unitary function is said to be an *inner function.*

In the first section we showed how the structure of contractions could be reduced to the study of their associated minimal unitary dilations and that this in turn reduced to the study of nested pairs of invariant subspaces for the unitary dilation. In the preceding paragraph we indicated that with certain unitary operators and certain nested pairs of invariant subspaces we could associate an analytic operator-valued function. Later we show that this is always possible and that this function is a complete set of unitary invariants for the contraction. This function is called the "characteristic operator function."

The characteristic operator function always determines a pair of wandering subspaces for a certain unitary operator. If the characteristic operator function is a unitary function, then we could already describe the procedure for constructing the wandering subspaces. However, one must also allow contractive functions and we want to indicate how a wandering subspace can be associated in this case.

Let C be a separable Hilbert space and Θ_λ be a contraction from C to E_λ for each λ in $\mathbf{T}$ such that $\Theta_\lambda x$ lies in $\int_{\mathbf{T}} \bigoplus E_\lambda \, d\lambda/2\pi$ for each x in C. Now unless Θ_λ is a partial isometry with a fixed kernel, the manifold ΘC will not be a wandering subspace for M_z. In general, ΘC will not be closed. However, there is a more or less natural way in which to enlarge the space and to obtain a wandering subspace for the extended unitary operator. Let Δ_λ denote the nonnegative square root of the operator $I - \Theta_\lambda^* \Theta_\lambda$ which acts on C and let $\bar{\Delta}_\lambda$ denote the projection onto the closure of the range of Δ_λ. The enlarged direct integral is formed by adding $\int_{\mathbf{T}} \bigoplus \bar{\Delta}_\lambda C \, d\lambda/2\pi$ to $\int_{\mathbf{T}} \bigoplus E_\lambda \, d\lambda/2\pi$. In this case we are interested in a particular maximal linear subspace to define $\int_{\mathbf{T}} \bigoplus \bar{\Delta}_\lambda C \, d\lambda$; namely the maximal linear subspace determined by $\{\varphi(\lambda)\Delta_\lambda x : \varphi \in L^\infty, x \in C\}$. We now consider the unitary function Θ'_λ defined from C to $\int_{\mathbf{T}} \bigoplus (E_\lambda \oplus \bar{\Delta}_\lambda C) d\lambda/2\pi$ such that $\Theta'_\lambda x = \Theta_\lambda x \oplus \Delta_\lambda x$. The wandering subspace defined by Θ' is the one in which we are interested.

Now there are certain other possibilities for this wandering subspace. If V_λ is a unitary function in $\int_{\mathbf{T}} \oplus L(\bar{\Delta}_\lambda C)\,d\lambda/2\pi$, then the operator function Ψ_V defined by $\Psi_V x = \Theta_\lambda x \oplus V_\lambda \Delta_\lambda x$ yields a wandering subspace which has the same "trace" on $\int_{\mathbf{T}} \oplus E_\lambda\,d\lambda/2\pi$ as that defined by Θ'. In most instances the wandering subspace defined by Θ' is all that is required and shall suffice for our purposes. It should be recalled, however, that there are these other possibilities.

3. Absolute continuity and the functional calculus. Earlier we showed that every contraction is obtained by compressing a unitary operator to a semi-invariant subspace. We then proceeded to obtain various representation theorems for unitary operators and their invariant subspaces. We now apply these results to show that the unitary dilation of a completely nonunitary contraction has an absolutely continuous scalar spectral measure or more succinctly, is *absolutely continuous.*

If W is a unitary operator on the Hilbert space K such that H is a full semi-invariant subspace for W, then $T = P_H W|H$ is a contraction operator on H having minimal unitary dilation W on K. By Corollary 1.4 in the first section, there exist invariant subspaces G and G_* for W and W^*, respectively, such that $K = G_* \oplus H \oplus G$. If G were not a pure invariant subspace for W, then G would contain a reducing subspace R for W. Since R would be contained in $H^\perp$, this would contradict the assumption that H is full. Similarly, G_* must be a pure invariant subspace for W^*. Thus we see that the semi-invariant subspaces can be described entirely in terms of the pure invariant subspaces G and G_* for W and W^*, respectively.

The following result indicates a further relation between T and the pair G and G_*.

LEMMA 3.1. *The contraction T on H is completely nonunitary if and only if the smallest reducing subspace for W containing G and G_* is K.*

PROOF. If Q denotes the smallest reducing subspace for W containing G and G_*, then $Q^\perp$ is contained in H. The result follows since $T|Q^\perp = W|Q^\perp$ is a unitary operator.

Now assume that W on K has the direct integral decomposition as M_z acting on $\int_{\mathbf{T}} \oplus E_\lambda\,d\nu$ and that T is completely nonunitary. If L and L_* are the wandering subspaces for G and G_* respectively, and D and D_* are Hilbert spaces of the same dimension, then there exist unitary functions Ψ and Ψ_* such that $L = \Psi D$ and $L_* = \Psi_* D_*$. Moreover, the smallest reducing subspaces for W containing G and G_* are therefore $\int_{\mathbf{T}} \oplus \Psi_\lambda(D)\,d\nu$ and

$\int_{\mathbf{T}} \oplus \Psi_{*\lambda}(\mathcal{D}_*)d\nu$, respectively. Thus, if F_λ denotes the subspace of E_λ spanned by $\Psi_\lambda(\mathcal{D})$ and $\Psi_{*\lambda}(\mathcal{D}_*)$, then the smallest reducing subspace for W containing both G and G_* is $\int_{\mathbf{T}} \oplus F_\lambda d\nu$ and hence is $\int_{\mathbf{T}} \oplus E_\lambda d\nu$ for completely nonunitary T by the lemma. Therefore, we see that the span of $\Psi_\lambda(\mathcal{D})$ and $\Psi_{*\lambda}(\mathcal{D}_*)$ is E_λ for almost all λ in $\mathbf{T}$. Since the measures $\dim \Psi_\lambda(\mathcal{D})d\nu(\lambda)$ and $\dim \Psi_{*\lambda}(\mathcal{D}_*)d\nu(\lambda)$ are both mutually absolutely continuous with respect to Lebesgue measure by remarks in the preceding section, we see that ν is mutually absolutely continuous with respect to Lebesgue measure. Thus we have proved

THEOREM 3. *The minimal unitary dilation for a completely nonunitary contraction is absolutely continuous.*

This result is due to Sz.-Nagy and Foiaş **[69]** who were motivated by the earlier result of Schreiber **[63]** which stated that the minimal unitary dilation of a strict contraction $(\|\cdot\| < 1)$ is a bilateral shift. Schreiber was interested in constructing an extended functional calculus **[64]** for strict contractions. We shall also be interested in such a functional calculus after we derive certain ancillary results.

It is convenient to introduce a classification of contractions at this point. A contraction T on H is said to belong to

class $C_{0.}$ if $\lim_{n\to\infty} T^n f = 0$ for f in H;
class $C_{.0}$ if $\lim_{n\to\infty} T^{*n} f = 0$ for f in H;
class $C_{1.}$ if $\lim_{n\to\infty} \|T^n f\| > 0$ for f in H; and
class $C_{.1}$ if $\lim_{n\to\infty} \|T^{*n} f\| > 0$ for f in H.

All combinations are possible and we set $C_{\alpha\beta} = C_{\alpha.} \cap C_{.\beta}$ for $\alpha, \beta = 0, 1$. This classification is due to Sz.-Nagy and Foiaş.

A contraction T seldom belongs to one of these classes, since, in general, there will be some vectors f for which $\lim_{n\to\infty} T^n f = 0$ and some for which it does not. The collection of all vectors for which this limit is zero forms an invariant subspace for T and T restricted to this subspace belongs to class $C_{0.}$. Proceeding in this manner one can attempt to reduce questions concerning arbitrary contractions to the extreme cases which these classes represent.

Before proceeding let us consider how this classification applies to some well-known operators. If T is a normal contraction on the Hilbert space H, then this classification can be described using the spectral measure $E(\Delta)$ for T. More precisely, one can see that $\lim_{n\to\infty} T^n f = 0$ if and only if f lies in $E(\mathbf{D})H$ and that $\lim_{n\to\infty} \|T^n f\| > 0$ if and only if f lies in $E(\mathbf{T})H$. Thus, we can write $T = T_1 \oplus T_2$, where T_1 and T_2 lie in classes C_{00} and C_{11}, respectively. In fact, T_2 is unitary.

A slightly more complicated but more typical example is obtained by considering bilateral weighted shifts. Let $\{e_n\}_{n=-\infty}^{\infty}$ be the canonical orthonormal basis for $l^2(\mathbf{Z})$.

An operator S is said to be a *bilateral weighted shift* relative to this basis if there exists a bounded sequence of nonnegative weights $\{\lambda_n\}_{n \in \mathbf{Z}}$ such that

$$S\left(\sum_{n=-\infty}^{\infty} \alpha_n e_n\right) = \sum_{n=-\infty}^{\infty} \alpha_{n-1}\lambda_n e_n.$$

The operator S is a contraction if $|\lambda_n| \leqslant 1$ for n in $\mathbf{Z}$ and elementary computations show that S lies in class $C_{0.}$ if the product $\Pi_{n=0}^{\infty} \lambda_n$ diverges to 0 and in class $C_{1.}$ if it does not. Similarly, S belongs to class $C_{.0}$ if $\Pi_{n=-\infty}^{0} \lambda_n$ diverges to 0 and to $C_{.1}$ if it does not.

The relation of this classification to canonical models is contained in

PROPOSITION 3.2. *The wandering subspace* L $[L_*]$ *is complete for* W *if and only if the operator* T *is in class* $C_{0.}$ $[C_{.0}]$. *In particular,* W *is a bilateral shift if* T *belongs to either class* $C_{0.}$ *or* $C_{.0}$.

PROOF. Firstly, the isometric isomorphism Θ from $\mathbf{H}^2_D$ onto G can be extended to an isometric isomorphism, also denoted Θ, between $\mathbf{L}^2_D$ and the smallest reducing subspace P for W containing G such that $\Theta U = W\Theta$. Secondly, if f is a vector in H, then $T^n f = P_H W^n f = (I - P_G)W^n f$, since $H \oplus G$ is invariant for W.

Thus, if L is complete we have for all f in H that

$$\lim_{n\to\infty} \|T^n f\| = \lim_{n\to\infty} \|(I - P_G)W^n f\| = \lim_{n\to\infty} \|P_{\mathbf{L}^2_D \ominus \mathbf{H}^2_D} U^n \Theta^* f\| = 0$$

and hence the sequence $\{T^n\}_{n=0}^{\infty}$ converges strongly to 0. Conversely, if $\lim_{n\to\infty} \|T^n f\| = 0$ for some f in H, then there exists a sequence $\{g_n\}_{n=0}^{\infty}$ in G such that $\lim_{n\to\infty} \|W^n f - g_n\| = 0$. Therefore, $\lim_{n\to\infty} \|f - W^{*n} g_n\| = 0$ and hence it follows that f is in the smallest reducing subspace for W containing G, and thus L is complete.

COROLLARY 3.3. *A contraction* T *is in class* C_{00} *if and only if each of the wandering subspaces* L and L_* *is complete for* W.

If T is in class C_{00}, then we have $H \oplus \mathbf{H}^2_L = \mathbf{H}^2_{L_*}$, and $\mathbf{H}^2_L$ is a full range invariant subspace for W contained in $\mathbf{H}^2_{L_*}$. Thus there exists an analytic inner function Θ such that each $\Theta(e^{it})$ is a unitary operator from L_* onto L and $\mathbf{H}^2_L = \Theta \mathbf{H}^2_{L_*}$. In this case Θ is the characteristic operator function for

T. Moreover, if one is given such an inner function, it is clear how to construct a contraction having Θ as its characteristic operator function. We shall have more to say about this later.

The preceding proposition enables us to obtain Schreiber's result as a corollary.

COROLLARY 3.4. *The minimal unitary dilation for a strict contraction is a bilateral shift.*

This corollary raises the question of whether every unitary dilation is a bilateral shift or more precisely which unitary operators occur as the minimal unitary dilation for a completely nonunitary contraction. To see that there are some restrictions we argue as follows. Recalling the notation of the proof of Theorem 2 and the fact that E_λ is the span of $\Psi_\lambda(\mathcal{D})$ and $\Psi_{*\lambda}(\mathcal{D}_*)$ for almost all λ, we see that

$$\max\ \{d, d_*\} \leqslant \dim E_\lambda \leqslant d + d_*,$$

where $d = \dim \mathcal{D}$ and $d_* = \dim \mathcal{D}_*$. This inequality already rules out certain dimension functions; for example, if E_λ is finite dimensional on a set of positive measure, then both d and d_* must be finite and hence it follows that $\dim E_\lambda < \infty$ a.e. Alternately, if $\dim E_\lambda = \infty$ on a set of positive measure, then W is a bilateral shift.

One can show that the preceding inequality is the only restriction on the dimension function. We shall do this after we give a more direct method of determining the cardinal numbers d and d_*.

If T is a contraction on H, recall that $D_T = (I - T^*T)^{1/2}$ and $D_{T^*} = (I - TT^*)^{1/2}$. These operators are called the *defect operators* for T and are zero only for unitary T. Moreover, there operators figure prominently in various geometric constructions of the unitary dilation for T (cf. **[20]**, **[73]**).

PROPOSITION 3.5. *If T is a contraction on H, then the rank of D_T $[D_{T^*}]$ is equal to the dimension d $[d_*]$ of the wandering subspace L $[L_*]$ for the unitary dilation W. Consequently, if* $\dim E_\lambda$ *is the multiplicity function for W, then* $\max\ \{d, d_*\} \leqslant \dim E_\lambda \leqslant d + d_*$ *a.e.*

PROOF. By our preceding remarks it is sufficient to show that the new definition of d agrees with the old one.

As we have previously remarked, $H \oplus G$ is an invariant subspace for W. If we write $G = \Sigma_{n=0}^{\infty} \bigoplus W^n L$, then we have $H \oplus L \oplus WL \oplus W^2 L \oplus \cdots$ is an invariant subspace for W. If f is a vector in H and g is a vector in G, then f and g are orthogonal. Thus f is orthogonal to Wg which implies that Wf lies in $\mathcal{H} \oplus \mathcal{L}$. If A is the operator defined from $\mathcal{H}$ to $\mathcal{L}$ such that $Wf = Tg \oplus Af$ for f in $\mathcal{H}$, then

$$\|f\|^2 = \|Wf\|^2 = \|Tf \oplus Af\|^2 = \|Tf\|^2 + \|Af\|^2$$

and hence

$$\|Af\|^2 = \|f\|^2 - \|Tf\|^2 = (f, f) - (T^*Tf, f) = \|D_T f\|^2.$$

Therefore, D_T is unitarily equivalent to $A: \mathcal{H} \longrightarrow \mathcal{L}'$, where $\mathcal{L}' = \text{clos } [A\mathcal{H}]$. However, since $\mathcal{G}_* \oplus \mathcal{H} \oplus \Sigma_{n=0}^{\infty} \bigoplus W^n \mathcal{L}'$ is obviously a reducing subspace for W containing $\mathcal{H}$, we see that $\mathcal{L} = \mathcal{L}'$ and thus the result follows.

Combining this characterization of defect numbers with Corollary 3.3 yields.

COROLLARY 3.6. *If T is a contraction in class C_{00}, then $d = d_*$.*

Since an absolutely continuous unitary operator with constant infinite multiplicity is a bilateral shift we have from Proposition 3.5 that

COROLLARY 3.7. *If T is a contraction operator for which either d or d_* is infinite, then its unitary dilation is a bilateral shift of infinite multiplicity.*

From this we obtain the affirmative answer due to Sz.-Nagy and Foiaş of a question raised by de Bruijn **[13]**.

COROLLARY 3.8. *If T is a contraction operator on the infinite-dimensional Hilbert space $\mathcal{H}$ which satisfies $\|Tf\| < \|f\|$ for every nonzero f in $\mathcal{H}$, then the unitary dilation is a bilateral shift of infinite multiplicity.*

PROOF. Since the hypothesis implies that T is completely nonunitary and the selfadjoint operator $(I - T^*T)^{1/2}$ has no kernel, then $d = \dim \mathcal{H}$ and the result follows.

If $\mathcal{H}$ is finite dimensional, then the assumption that $\|Tf\| < \|f\|$ for every nonzero f in $\mathcal{H}$ implies that T is a strict contraction and hence the unitary dilation is a bilateral shift by Corollary 3.4.

Since we have shown in most of the previous results that the unitary dilation is a bilateral shift under rather diverse circumstances, a simple example when it is not might be of interest.

Let φ be a bounded holomorphic function on $\mathbf{D}$ having the boundary function $\hat{\varphi}$ defined on $\mathbf{T}$ which satisfies

$$|\hat{\varphi}(e^{it})| = \begin{cases} 3/5 & \text{for } 0 \leqslant t < \pi, \\ 1 & \text{for } \pi \leqslant t < 2\pi. \end{cases}$$

We set

$$\Delta(e^{it}) = \begin{cases} 4/5 & \text{for } 0 \leqslant t < \pi \\ 0 & \text{for } \pi \leqslant t < 2\pi, \end{cases}$$

and let $\bar{\Delta}$ denote the characteristic function of the upper half circle. We shall produce a full semi-invariant subspace for M_z acting on

$$\int_{\mathbf{T}} \oplus E_\lambda \frac{d\lambda}{2\pi}, \quad \text{where } E_{e^{it}} = \begin{cases} \mathbf{C}^2 & \text{for } 0 \leqslant t < \pi, \\ \mathbf{C} & \text{for } \pi \leqslant t < 2\pi. \end{cases}$$

We let $\{x_1, x_2\}$ denote an orthonormal basis for $\mathbf{C}^2$ with x_1 in $\mathbf{C}$, and then consider the wandering subspaces $L_* = \Psi_*\mathbf{C}$ and $L = \Psi\mathbf{C}$, where $\Psi_*(e^{it}) = e^{it}x_1$, and $\Psi(e^{it}) = \hat{\varphi}(e^{it})x_1 + \Delta(e^{it})x_2$. It is easy to check that L and L_* are wandering subspaces and that

$$G = \sum_{n=0}^{\infty} \oplus M_z^n L = \{\varphi(e^{it})f(e^{it})x_1 + \Delta(e^{it})f(e^{it})x_2 : f \in H^2\}$$

and

$$G_* = \sum_{n=0}^{\infty} \oplus M_z^{*n} L_* = \{g(e^{it})x_1 : g \in K^2\}$$

are pairwise orthogonal. Moreover, since the smallest reducing subspace for M_z containing G and G_* is clearly $\int_{\mathbf{T}} \oplus E_\lambda\, d\lambda/2\pi$, we see that

$$H = \int_{\mathbf{T}} \oplus E_\lambda\, d\lambda/2\pi \ominus \{G_* \oplus G\}$$

is a full semi-invariant subspace for M_z such that $T = P_H M_z | H$ is a completely nonunitary contraction. Lastly, since the multiplicity function for M_z is not constant, M_z is not a bilateral shift.

This example contains the essential ingredients for completing the characterization of those unitary operators which can be the unitary dilation of a completely nonunitary contraction.

PROPOSITION 3.9. *If W is an absolutely continuous unitary operator on K having multiplicity function $\eta(\lambda)$, then there exists a full semi-invariant subspace H for W if and only if*

$$\eta(\lambda) \leqslant 2 \min_{\mu \in \mathbf{T}} \eta(\mu) \quad \textit{for } \lambda \textit{ in } \mathbf{T}.$$

PROOF. An easy argument pointed out to the author by Arveson shows that $\eta(\lambda) \leqslant 2 \min_{\mu \in \mathbf{T}} \eta(\mu)$ if and only if there exist cardinals c and c_* such that $\max\{c, c_*\} \leqslant \eta(\lambda) \leqslant c + c_*$. (Take $c = \min_{\mu \in \mathbf{T}} \eta(\mu)$ and $c_* = \max_{\mu \in \mathbf{T}} \eta(\mu)$.)

Thus, we have established one half and hence may assume the existence of such c and c_*. Moreover, if either c or c_* is infinite, then W is unitarily equivalent to the bilateral shift U on $\mathbf{L}^2_{\mathcal{D}}$, where $\mathcal{D}$ is infinite dimensional. In this case, the constant subspace $\mathcal{D}$ has the required properties.

Otherwise assume that c and c_* are finite and that $c_* \leqslant c$. Set $\Delta_j = \{\lambda \in \mathbf{T}: \eta(\lambda) = c + j\}$ for $1 \leqslant j \leqslant c_*$. Then $\{\Delta_j\}_{j=1}^{c*}$ is a partition of $\mathbf{T}$. For $1 \leqslant i \leqslant c$ let φ_i be a bounded holomorphic function on $\mathbf{D}$ such that $|\varphi_i(z)| \leqslant 1$ for z in $\mathbf{D}$ and having a boundary function $|\hat{\varphi}_i(\lambda)| < 1$ for λ in $\bigcup_{i \leqslant j \leqslant c_*} \Delta_j$. Consider the direct integral defined with scalar spectral measure $d\theta/2\pi$ and $E_\lambda = \mathbf{C}^{\eta(\lambda)}$. Then the unitary function Φ_* defined from $\mathbf{C}^c$ to $E_\lambda = \mathbf{C}^c \oplus \mathbf{C}^{\eta(\lambda)-c}$ by $\Phi_{*e^{i\theta}}(x) = e^{-\theta}x \oplus 0$ gives a wandering subspace $\Phi_*\mathbf{C}^c$ for M_z. Further, the unitary function Φ defined from $\mathbf{C}^c$ to E_λ by

$$\Phi_{e^{i\theta}}(x_1, x_2, \cdots, x_c) = (\hat{\varphi}_1(e^{i\theta})x_1, \hat{\varphi}_2(e^{i\theta})x_2, \cdots, \hat{\varphi}_c(e^{i\theta})x_c,$$
$$\psi_1(e^{i\theta})x_1, \cdots, \psi_{\eta(\lambda)-c}(e^{i\theta})x_{\eta(\lambda)-c})$$

where $\psi_i(e^{i\theta}) = (1 - |\hat{\varphi}_i(e^{i\theta})|^2)^{1/2}$ for $1 \leqslant i \leqslant c_*$ also gives a wandering subspace $\Phi\mathbf{C}^c$ for M_z. Moreover, the corresponding invariant subspaces

$$G = \sum_{n=0}^{\infty} \oplus M_z^n \Phi \mathbf{C}^c \quad \text{and} \quad G_* = \sum_{n=-\infty}^{0} \oplus M_z^n \Phi_* \mathbf{C}^c$$

for M_z and M_z^* are pairwise orthogonal, and $\int_{\mathbf{T}} \oplus E_\lambda \, d\theta/2\pi \ominus (G_* \oplus G)$ is the desired semi-invariant subspace.

Although we have proved the theorem, there remains one loose end. If c is finite and c_* infinite, then a closed subspace $\mathcal{D}$ of dimension c of the Hilbert space E of dimension c_* can be used to define the semi-invariant subspace $\mathcal{D} \oplus \mathbf{H}^2_{E \ominus \mathcal{D}} = F$ for U on $\mathbf{L}^2_E$, and the contraction $T = P_F W | F$ has defect numbers $d = c$ and $d_* = c_*$, respectively. Thus a semi-invariant subspace with the right defect numbers exists. A similar construction works when c_* is finite but c is infinite.

We now consider the refined functional calculus for contractions due to Sz.-Nagy and Foiaş. If T is a contraction on H and φ is a function holomorphic on a neighborhood of the closed unit disk $\overline{\mathbf{D}}$, then following F. Riesz one defines

$$\varphi(T) = \frac{1}{2\pi i} \int_{\mathbf{T}} (T - \lambda)^{-1} \, d\lambda.$$

If $\varphi(z) = \Sigma_{n=0}^{\infty} a_n z^n$ is the Taylor series expansion for φ, then it is easily shown that $\varphi(T) = \Sigma_{n=0}^{\infty} a_n T^n$, where the later series converges uniformly in the norm topology. Moreover, the map $\varphi \longrightarrow \varphi(T)$ defines a homomorphism from the algebra of all holomorphic functions on some neighborhood of $\bar{\mathbf{D}}$ into $L(H)$. The importance of this functional calculus is lessened, however, by the limited nature of the functions which are allowed.

If W is a unitary operator on K with the direct integral decomposition M_z on $\int_{\mathbf{T}} \oplus E_\lambda \, d\nu$, then a functional calculus can be defined for W based on the algebra $L^\infty(\nu)$ such that $\varphi(W)$ corresponds to M_φ. If W is the unitary dilation for a contraction T, then it is natural to investigate the mapping $\varphi \longrightarrow P_H \varphi(W)|H$. A moment's reflection, however, will reveal that this mapping is seldom a homomorphism. The difficulty arises already with the functions z and $\bar{z}$. Thus one is led to consider the restriction of this mapping to the "closed" subalgebra of $L^\infty(\nu)$ generated by z. Although the right notion of closure is not obvious, one clearly wants the closure as large as possible, or, equivalently, the closure in as weak a topology on $L^\infty(\nu)$ as possible. Rather than proceeding further in this direction let us restrict our attention to completely nonunitary contractions and start anew.

If T is a completely nonunitary contraction on H, then there is a direct integral representation for the unitary dilation W in which M_z acts on $\int_{\mathbf{T}} \oplus E_\lambda d\lambda/2\pi$ by Theorem 3. If π denotes the mapping defined by $\pi(\varphi) = M_\varphi$ from $L^\infty(d\lambda/2\pi)$ to $L(\int_{\mathbf{T}} \oplus E_\lambda \, d\lambda/2\pi)$, then, as observed in **[61]**, π is not only continuous in the norm topology but is, in fact, continuous as a mapping for the weak *-topology on $L^\infty(d\lambda/2\pi)$ and the weak operator topology on $L(\int_{\mathbf{T}} \oplus E_\lambda \, d\lambda/2\pi)$. The smallest weak *-closed subalgebra of $L^\infty(d\lambda/2\pi)$ containing the function z is the Hardy algebra H^∞. This algebra has an alternate characterization as the bounded holomorphic functions on $\mathbf{D}$. If ψ is a bounded holomorphic function on $\mathbf{D}$ then a lemma due to Fatou asserts that $\lim_{r \to 1^-} \psi(re^{it})$ exists a.e. and the resulting function $\hat{\psi}$ defined on $\mathbf{T}$ can be shown to belong to H^∞. We shall not prove this but refer the reader to the excellent books of Hoffman **[42]** and Duren **[26]** on this subject.

We can now state our main result on the functional calculus for contractions.

THEOREM 4. *If T is a completely nonunitary contraction on H with minimal unitary dilation W on K, then the functional calculus defined by $\varphi(T) = P_H \varphi(W)|H$ for φ in H^∞ is a contractive homomorphism from H^∞ into $L(H)$ which extends the Riesz functional calculus. Moreover, this mapping is continuous from the weak *-topology on H^∞ to the weak operator topology on $L(H)$.*

PROOF. The continuity of this mapping in the "weak topologies" follows from our previous remarks. That the mapping is contractive follows from the inequality $\|\varphi(T)\| \leqslant \|\varphi(W)\| \leqslant \|\varphi\|_\infty$, where $\|\varphi\|_\infty$ denotes the essential supremum of φ on $\mathbf{T}$. The agreement of $\varphi(T)$ with the usual definition when φ is a polynomial in z follows from the fact that $p(T) = P_H p(W)|H$. Thus this mapping is a homomorphism on polynomials and hence on all of H^∞ since the polynomials are weak *-dense in H^∞.

These properties of the functional calculus enable us to give a more constructive definition which is the primary one for Sz.-Nagy and Foiaş. If φ is a function in H^∞ then φ has a holomorphic extension $\hat{\varphi}$ to $\mathbf{D}$. If we define $\varphi_r(e^{it}) = \hat{\varphi}(re^{it})$ for $0 < r < 1$ and e^{it} in $\mathbf{T}$, then $\lim_{r\to 1-} \varphi_r = \varphi$ in the weak *-topology on H^∞ (cf. **[42]**). Thus we obtain, from Theorem 4,

COROLLARY 3.10. *If T is a completely nonunitary contraction on H and φ is a function in H^∞, then $\lim_{r\to 1-}\varphi_r(T)$ exists in the weak operator topology and equals $\varphi(T)$.*

The advantage of this definition is that φ_r is holomorphic on a neighborhood of the closed unit disk and hence $\varphi_r(T)$ can be defined using the Taylor series expansion for φ.

Actually, one can show that $\lim_{r\to 1-} \varphi_r(T) = \varphi(T)$ in the strong operator topology.

We conclude this section by defining a very tractable class of operators and then deriving certain preliminary results concerning them. If T is an operator defined on a finite-dimensional Hilbert space, then there is a polynomial $p(z)$ for which $p(T) = 0$. Most of the known results concerning such algebraic operators can be shown to follow from this. Thus it is natural to investigate operators T for which there exists some function φ such that $\varphi(T) = 0$. If φ is holomorphic on a neighborhood of the spectrum of T, then there is a smaller neighborhood on which φ has at most finitely many zeroes. If we write $\varphi(z) = p(z)\psi(z)$ where ψ does not vanish on this neighborhood, then $0 = \varphi(T) = p(T)\psi(T)$. Since ψ does not vanish on a neighborhood of the spectrum of T, one can define $(1/\psi)(T)$ and hence $\psi(T)$ is invertible. Thus the polynomial $p(z)$ annihilates T and no greater generality has been attained.

If T is a completely nonunitary contraction and φ is a nonzero function in H^∞, then the vanishing of $\varphi(T)$ need not imply the existence of a polynomial which annihilates T. There is an extensive theory due to Sz.-Nagy and Foiaş concerning this *class of* C_0*-operators.*

What operators are in class C_0? Obviously, the finite rank operators belong to the class C_0. Further, if φ is an inner function in H^∞ (that is, $|\varphi| = 1$ a.e.), then the operator S_φ obtained by compressing M_z to $H^2 \ominus \varphi H^2 = \mathcal{M}_\varphi$ lies in class C_0. To show this let f lie in H^2 and observe that φf is orthogonal to $\mathcal{M}_\varphi$. Hence $\varphi(S_\varphi)f = 0$ and therefore we have $\varphi(S_\varphi) = 0$, which implies that S_φ lies in class C_0.

Further, if T belongs to class C_0, then there is a function ψ in H^∞ such that $\psi(T) = 0$. By the previous corollary, if $\psi(z) = \Sigma_{n=0}^\infty a_n z^n$, then $\lim_{r \to 1-} \{\Sigma_{n=0}^\infty a_n r^n T^n\} = \psi(T) = 0$ in the weak operator topology. Thus we have $\tilde{\psi}(T^*) = 0$, where $\tilde{\psi}(z) = \Sigma_{n=0}^\infty \bar{a}_n z^n$, since $(\Sigma_{n=0}^\infty a_n r^n T^n)^* = (\Sigma_{n=0}^\infty \bar{a}_n r^n T^{*n})$. Therefore, T is in C_0 if and only if T^* is in C_0.

How does one use the fact that an operator lies in class C_0 to study it? In linear algebra one usually begins by establishing the existence of a minimal polynomial. One method of proof involves observing that the set of polynomials that annihilate T is an ideal and invoking the fact that the ring of polynomials is a principal ideal domain. A similar result is true about H^∞ and may be used to establish the existence of a minimum function for C_0-operators.

PROPOSITION 3.11. *Every weak *-closed ideal in* H^∞ *is of the form* φH^∞ *for some inner function* φ.

This result is due to Srinivasan and Wang (cf. **[38]**) and is closely related to Beurling's characterization of the invariant subspaces of multiplication by z on H^2. In fact, for weak *-closed subspaces of H^∞, the notions of ideal and invariant subspace coincide.

COROLLARY 3.12. *If* T *is a class* C_0*-operator on* $\mathcal{H}$, *then there is an inner function* m_T *determined up to a scalar multiple such that for* φ *in* H^∞ *we have* $\varphi(T) = 0$ *if and only if* m_T *divides* φ.

PROOF. The result follows from the preceding proposition after observing that the set $\{\varphi \in H^\infty : \varphi(T) = 0\}$ is a weak *-closed ideal in H^∞.

If T is a contraction on a finite-dimensional Hilbert space, then there exists a minimal polynomial $p_T(z)$ such that $p_T(T) = 0$. Since p_T lies in H^∞, the operator T lies in class C_0, and hence has a minimal function m_T by the preceding corollary. How are m_T and p_T related? If $\alpha_1, \alpha_2, \cdots, \alpha_k$ are the roots of $p_T(z)$ counted multiply, then $p_T(z) = \alpha_0 \Pi_{j=1}^k (z - \alpha_j)$ and

$$\psi(z) = \prod_{j=1}^{k} \frac{z - \alpha_j}{1 - \bar{\alpha}_j z}$$

is an inner function, actually a finite Blaschke product. Since both $\psi(z)/p_T(z)$ and $p_T(z)/\psi(z)$ are bounded and holomorphic on **D**, we see that $m_T(z) = \psi(z)$.

Observe that if m_T is the minimal function for a contraction T in class C_0, then $\widetilde{m}_T$ is the minimal function for T^*.

In the next section we shall relate the spectrum of T to m_T and certain classes of invariant subspaces for T to factoring m_T. For now we content ourselves with one structural result. First we need a lemma.

LEMMA 3.13. *If T is a completely nonunitary contraction on $\mathcal{H}$ with unitary dilation W on $\mathcal{K}$, then $\lim_{n\to\infty} W^{-n}T^nf$ exists for f in $\mathcal{H}$ and defines an operator L from $\mathcal{H}$ to $\mathcal{K}$. Moreover, we have $\varphi(W)L = L\varphi(T)$ for φ in H^∞.*

PROOF. If $0 \leqslant m \leqslant n$, and f is in $\mathcal{H}$, then

$$\begin{aligned}(W^{-n}T^nf, W^{-m}T^mf) &= (W^{m-n}T^nf, T^mf) = (T^{*(n-m)}T^nf, T^mf)\\ &= (T^{*n}T^nf, f) = \|T^nf\|^2\end{aligned}$$

and hence

$$\begin{aligned}&\|W^{-n}T^nf - W^{-m}T^mf\|^2\\ &\qquad = \|W^{-n}T^nf\|^2 + \|W^{-m}T^mf\|^2 - 2\,\mathrm{Re}\,(W^{-n}T^nf, W^{-m}T^mf)\\ &\qquad = \|T^nf\|^2 + \|T^mf\|^2 - 2\|T^nf\|^2 = \|T^mf\|^2 - \|T^nf\|^2.\end{aligned}$$

Since $\lim_{n\to\infty} \|T^nf\|^2$ exists (the sequence $\{\|T^nf\|\}_{n=0}^\infty$ is monotonically decreasing), we have that $Lf = \lim_{n\to\infty} W^{-n}T^nf$ defines a contraction from $\mathcal{H}$ to $\mathcal{K}$. Lastly, since $LT^m = W^mL$ for each $m \geqslant 0$, the identity $L\varphi(T) = \varphi(W)L$ follows from the weak *-density of the polynomials in H^∞ and the corresponding continuity of the functional calculus.

Although the operator L is important in the study of unitary dilations by Sz.-Nagy and Foiaş , our interest is confined to the following application.

PROPOSITION 3.14. *Every contraction in class C_0 is in class C_{00}.*

PROOF. If φ is a inner function in H^∞, T is a completely nonunitary contraction on $\mathcal{H}$, and f is a vector in $\mathcal{H}$ such that $\varphi(T)f = 0$, then $\varphi(W)Lf = L\varphi(T)f = 0$. Since $\varphi(W)$ is a unitary operator we see that

$Lf = 0$, and hence

$$\lim_{n\to\infty} \|T^n f\| = \lim_{n\to\infty} \|W'^{-n} T^n f\| = \|Lf\| = 0,$$

thus concluding the proof.

Since, as we shall indicate, there exists a very nice theory for contractions in class C_0, it would be fortunate if the inclusion $C_0 \subset C_{00}$ were in fact an equality. That this is not the case can be illustrated with a familiar example. Let T_n be a contraction operator defined on $\mathbf{C}^n$ with minimum polynomial $p_n(z) = z^{n-1}$. For example, let T_n be the operator defined by the matrix consisting of all zeroes except for a diagonal of ones just below the main diagonal. The operator $T = \Sigma_{n=1}^{\infty} \oplus T_n$ is a contraction and an easy computation shows that $\lim_{k\to\infty} T^k f = 0$ and $\lim_{k\to\infty} T^{*k} f = 0$ for f in $\Sigma_{n=1}^{\infty} \oplus \mathbf{C}^n$ and hence T is in class C_{00}. Suppose that φ is a function in H^∞ such that $\varphi(T) = 0$. Since $\varphi(T) = \Sigma_{n=1}^{\infty} \oplus \varphi(T_n)$, this implies $\varphi(T_n) = 0$ for every integer n. But the minimal function for T_n is z^{n-1}, which implies that z^{n-1} divides φ for every n. Therefore $\varphi \equiv 0$ and thus T does not belong to class C_0.

If T is a contraction in class C_0, then by the remark following Corollary 3.3, T has an inner function Θ which is unitary valued from $\mathfrak{L}_*$ onto $\mathfrak{L}$ and has a holomorphic extension as a contraction valued mapping from $\mathfrak{L}_*$ into $\mathfrak{L}$ on the unit disk. It would seem reasonable that some close relationship should exist between Θ and the minimal function m_T for T. What is more important, we would like necessary and sufficient conditions in terms of Θ for T to belong to class C_0. While none are known, one can show that if D_T is a trace class operator and T is in C_{00}, then T is in class C_0. We establish this result under the stronger hypothesis that D_T is finite rank.

PROPOSITION 3.15. *If T is a contraction in class C_{00} and D_T is of finite rank, then T is class C_0. Moreover, m_T divides the "determinant" δ_T of the characteristic operator function Θ_T.*

PROOF. Since $\dim \mathfrak{L} = \dim \mathfrak{L}_*$, we may choose orthonormal bases for $\mathfrak{L}_*$ and $\mathfrak{L}$ and express Θ as a holomorphic unitary valued matrix function. If Ψ is the matrix obtained by forming cofactors, that is, the (i, j)th entry of Ψ is the (i, j)th cofactor of Θ, then $\Theta(e^{it})\Psi(e^{it}) = \Psi(e^{it})\Theta(e^{it}) = \delta_T(e^{it})$ by Cramer's rule, where $\delta_T(e^{it})$ is the determinant of $\Theta(e^{it})$. Since Θ is unitary valued, we see that δ_T is an inner function. Moreover, the inclusion $\delta_T \mathbf{H}^2_{\mathfrak{L}} = \Theta\Psi\mathbf{H}^2_{\mathfrak{L}} \subset \Theta\mathbf{H}^2_{\mathfrak{L}_*}$ follows from the fact that Ψ is holomorphic, and hence $H = \mathbf{H}_L^2 \ominus \Theta\mathbf{H}_{L_*}^2 \subset \mathbf{H}_L^2 \ominus \delta_T\mathbf{H}_L^2$. If f is in $\mathbf{H}_L^2$, then $M_{\delta_T}f$ lies in $\delta_T\mathbf{H}_L^2$, and therefore $P_H M_{\delta_T} f = 0$. Thus $\delta_T(T) = 0$, which completes the proof.

Although δ_T depends on the choice of bases for the domain and range of Θ, an easy argument using the multiplicativity of determinants shows that changing bases only multiplies δ_T by a constant of modulus one. Further, observe that $\delta_{T^*} = \widetilde{\delta}_T$.

The extension of this result to contractions T for which D_T is a trace class operator depends on defining the determinant for operators of the form $I + S$, where S is a trace class operator (cf. **[32]**).

4. **Characteristic operator functions.** The basic problem in operator theory is that of unitary equivalence, that is, determining when two given operators are unitarily equivalent. Although solving the general problem is undoubtedly hopeless, solutions for certain classes of operators have been found, and it is reasonable to expect that "good" solutions exist for others. A solution usually takes the form of attaching certain "simpler" objects to an operator called unitary invariants. For example, for a normal operator, a complete set of unitary invariants consists of a scalar spectral measure along with a multiplicity function. (See the article by A. Brown in this volume.)

Since there is little, if any, selfadjoint character to most operators, it is more natural to try to associate "analytic" unitary invariants to more general operators as opposed to "measure-theoretic" ones. Further, this propensity is strengthened by the analytic nature of the resolvent of an operator. One complete set of "analytic" unitary invariants is the "characteristic operator function." The characteristic operator function for a contraction operator is a holomorphic contraction valued function on $\mathbf{D}$ from which it is possible to construct the nested invariant subspaces for the minimal unitary dilation and hence the semi-invariant subspace which defines the contraction. We have already encountered a certain class (the C_{00}-class) of contractions for which we have defined the characteristic operator function. Unfortunately, in the general case the characteristic operator function is neither this simple nor as near the surface.

Assume that W is the unitary operator M_z acting on $K = \int_{\mathbf{T}} \oplus E_\lambda \, d\lambda/2\pi$ and that H is a full semi-invariant subspace with the representation

$$K = G_* \oplus H \oplus G,$$

where G_* and G are invariant for W^* and W, respectively, and that T is the compression of W to H. There are natural maps between $\mathbf{K}_{L_*}^2$ and G_* and $\mathbf{H}_L^2$ and G which we use to identify these pairs of spaces. Thus we obtain natural embeddings of $\mathbf{L}_{L_*}^2$ and $\mathbf{L}_L^2$ into $\int_{\mathbf{T}} \oplus E_\lambda \, d\lambda/2\pi$ as reducing subspaces for M_z. Therefore, we have an embedding of both L_* and L into E_λ for each λ in $\mathbf{T}$, and E_λ is the span of these two subspaces. If we denote the images of L_* and L by $L_{*\lambda}$ and L_λ, respectively, then $L_{*\lambda}$ and L_λ are, in general, not orthogonal. In fact, it is an amusing exercise to show that $L_{*\lambda} \perp L_\lambda$ a.e., if and only if the contraction T is the direct sum of a unilateral shift with the adjoint of a unilateral shift.

Now by Theorem 2 there exists an isometry Ψ from $\mathbf{H}^2_L$ into $\int_{\mathbf{T}} \bigoplus E_\lambda \, d\lambda/2\pi$ such that $\Psi\mathbf{H}^2_L = G$ and $\Psi U_+ = M_z\Psi$. We obtain the characteristic operator function by analyzing Ψ. In particular, since $\mathbf{L}^2_{L_*}$ is embedded as a reducing subspace for M_z, it follows that the projection valued function $P_{L_{*\lambda}}$ is a range function and hence lies in $\int_{\mathbf{T}} \bigoplus L(E_\lambda) d\lambda/2\pi$. Thus if we write $\Psi_\lambda = \Theta_\lambda \oplus \Phi_\lambda$, where

$$\Theta_\lambda = P_{L_{*\lambda}} \Psi_\lambda \quad \text{and} \quad \Phi_\lambda = P_{E_\lambda \ominus L_{*\lambda}} \Psi_\lambda,$$

then $\Theta\mathbf{H}^2_L$ is contained in $\mathbf{L}^2_{L_*}$, and $\Phi\mathbf{H}^2_L$ lies in $\int_{\mathbf{T}} \bigoplus (E_\lambda \ominus L_{*\lambda}) d\lambda/2\pi$.

Essentially, Θ is the characteristic operator function for T. Before explaining this remark further, let us record certain properties of Θ. Firstly, Θ is a contraction valued function from L to L_* which has a holomorphic extension to the interior of $\mathbf{D}$. The last part of this statement follows from the fact that $\Theta\mathbf{H}^2_L$ is contained in $\mathbf{H}^2_{L_*}$. To see this, recall that for f in $\mathbf{H}^2_L$ the vector $\Theta f \oplus \Phi f$ lies in G which is orthogonal to $\mathbf{K}^2_{L_*}$, thus implying that Θf is orthogonal to $\mathbf{K}^2_{L_*}$ and hence lies in $\mathbf{H}^2_{L_*}$. From the maximum modulus theorem we can deduce that $\Theta(z)$ is a contraction for z in $\mathbf{D}$. Lastly, we claim that Θ enjoys one further property; namely, $\|\Theta(0)x\| < \|x\|$ for each nonzero vector x in L. Suppose x is a vector in L for which

$$\|\Theta(0)x\| = \|x\|.$$

The vector-valued maximum modulus theorem yields that $\Theta(z)x = y$ for all z in $\mathbf{D}$ and hence the subspace $\{f(\cdot)y: f \in H^2\}$ is contained in G. Since $\{f(\cdot)y: f \in K^2\}$ is contained in G_*, it follows that the reducing subspace $\{f(\cdot)y: f \in L^2\}$ is contained in $H^\perp$, which contradicts the fact that H is full.

The preceding properties define the notion of characteristic operator function. A *characteristic operator function* is a holomorphic contraction-valued mapping $\Theta(z)$ on $\mathbf{D}$ between Hilbert spaces L and L_* which satisfies

$$\|\Theta(0)x\| < \|x\|$$

for each nonzero vector x in L.

A holomorphic contraction-valued function on $\mathbf{D}$ having this latter property is said to be *purely contractive* and an arbitrary holomorphic contraction-valued function $\{C(z), E, E_*\}$ decomposes into the direct sum $C(z) = D(z) \oplus V$, where $D(z)$ is a purely contractive holomorphic contraction-valued function from F to F_* and V is an isometric isomorphism from $E \ominus F$ onto $E_* \ominus F_*$.

We have stated that the characteristic operator function forms a complete set of unitary invariants, but that is not obvious. On the other hand it is clear that the unitary function $\Psi(\lambda)$ from L onto $E_\lambda = L_{*\lambda} \oplus (E_\lambda \ominus L_{*\lambda})$ does. If T lies in class $C_{.0}$ then $E_\lambda = L_{*\lambda}$, $\Psi = \Theta$, and in this case the characteristic operator coincides with the unitary function which locates G in $\mathbf{L}^2_{L_*}$. If T does not lie in class $C_{.0}$, then Θ is not a unitary function and is only contraction-valued. At the conclusion of the second section we indicated a natural way of associating a wandering subspace for a unitary operator on an enlarged Hilbert space for contraction operator functions. We want to show that this is exactly the relation between Θ and the defining unitary function.

Recall that $\Delta_\lambda = (I - \Theta_\lambda^*\Theta_\lambda)^{1/2}$ and $\bar{\Delta}_\lambda$ is the projection onto the closure of the range of Δ_λ. Since $\|\Phi f\|^2 = \|f\|^2 - \|\Theta f\|^2 = \|\Delta f\|^2$ for f in $\mathbf{H}^2_L$, it follows that there exists an isometry V from the closure of the range of Δ in $\int_{\mathbf{T}} \bigoplus \bar{\Delta}_\lambda L \, d\lambda/2\pi$ to the closure of the range of Φ in $\int_{\mathbf{T}} \bigoplus (E_\lambda \ominus L_{*\lambda}) d\lambda/2\pi$. Moreover, the operator V intertwines M_z on both spaces, hence extends to the corresponding reducing subspaces and therefore can be defined by an isometric isomorphism from $\bar{\Delta}_\lambda L$ onto $E_\lambda \ominus L_{*\lambda}$. Thus if we consider M_z acting on the direct integral $\int_{\mathbf{T}} \bigoplus (L_* \oplus \bar{\Delta}_\lambda L) d\lambda/2\pi$, then G_* is identified with $\mathbf{K}^2_{L_*}$ and G with $\{\Theta f \oplus \Delta f : f \in \mathbf{H}^2_L\}$. The preceding recipe works for a general characteristic operator function and we state the result as

THEOREM 4. *If $\{\Theta, L, L_*\}$ is a characteristic operator function, then*

$$H = \left[\mathbf{H}^2_{L_*} \oplus \int_{\mathbf{T}} \bigoplus \bar{\Delta}_\lambda L \frac{d\lambda}{2\pi}\right] \ominus \{\Theta f \oplus \Delta f : f \in \mathbf{H}^2_L\}$$

is a full semi-invariant subspace for the unitary operator M_z acting on $\int_{\mathbf{T}} \bigoplus (L_* \oplus \bar{\Delta}_\lambda L) d\lambda/2\pi$. If $T = P_H M_z | H$, then T is a completely nonunitary contraction with minimal unitary dilation M_z. Moreover, for every completely nonunitary contraction T, there exists such a characteristic operator function $\{\Theta_T, L, L_*\}$. Further, if $\{\Theta, L, L_*\}$ and $\{\Theta', L', L'_*\}$ are characteristic operator functions, then the corresponding contractions which they define are unitarily equivalent if and only if there exist isometric isomorphism η and η_* from L onto L' and L_* onto L'_*, respectively, such that

$$\eta_*\Theta(z) = \Theta'(z)\eta \quad \text{for } z \text{ in } \mathbf{D}.$$

Although we have not proved all the parts of this result in detail, we have given all the basic ideas and filling in the missing steps would provide a challenging exercise for the reader (cf. [73]).

If one attempts to define a canonical model for a holomorphic contraction-valued function on **D** which is not purely contractive, then the unitary dilation would not be minimal.

The notion of characteristic operator function occurs first in the work of Livšic [**50**]. Our usage coincides with that of Sz.-Nagy and Foiaş, although our definition and derivation are somewhat different. In particualr, they obtain the following explicit formula for Θ_T.

PROPOSITION 4.1. *If* T *is a completely nonunitary contraction on* $\mathcal{H}$, $\mathcal{D}_T = \text{clos}\,[D_T\mathcal{H}]$ *and* $\mathcal{D}_{T*} = \text{clos}\,[D_{T*}\mathcal{H}]$, *then the operator function defined by*

$$\Theta_T(z) = -T + \sum_{n=1}^{\infty} z^n D_{T*} T^{*n-1} D_T$$

for z *in* **D** *and mapping* $\mathcal{D}_T$ *into* $\mathcal{D}_{T*}$ *is a characteristic operator function for* T.

That $\Theta_T(z)\mathcal{D}_T$ is actually contained in $\mathcal{D}_{T*}$ follows from the identity $TD_T = D_{T*}T$, proved in the first section. Although we do not prove this proposition, a proof could be given working in our context by first establishing for a given characteristic operator function that $\Theta(0)$ is unitarily equivalent to $\Theta_T(0)$ and then extending this result to z in **D** by considering the contraction $T_z = (T-z)(I-\bar{z}T)^{-1}$. (One can show that $\Theta_{T_z}(0)$ is unitarily equivalent to $\Theta_T(z)$ (cf. [73, p. 240]).) Such an approach might be important for canonical models defined over regions more complicated than the disk.

Observe that it follows from the proposition that the characteristic operator function Θ_{T*} for T^* satisfies $\Theta_{T*}(z) = \Theta_T(\bar{z})^*$.

We began this section with a discussion of the unitary equivalence problem; theoretically the characteristic operator function provides a solution to this problem. Two contractions are unitarily equivalent if and only if their characteristic operator functions are equivalent. In most cases, however, one can hardly call this a solution.

If both defect numbers are finite, however, then a reduction has been made since the study of an infinite-dimensional operator has been replaced by the study of a holomorphic matrix function. We conclude these philosophical comments by reminding the reader that even in this latter case a great many questions remain.

Since its characteristic operator function forms a complete set of unitary invariants for a contraction, it is at least theoretically possible to deduce all properties of the contraction from corresponding properties of the characteristic operator function. In the remainder of this section we shall attempt to interpret the spectrum and the invariant subspaces for the contraction in terms of the characteristic operator function.

We begin with the spectrum; although we state the result precisely, we will only sketch the proof.

Since T is a contraction, the spectrum of T is contained in the closed unit disk. The question of whether a z in $\bar{\mathbf{D}}$ lies in $\sigma(T)$ is conveniently divided into two cases. If z_0 is an interior point, then the invertibility of $T - z_0$ can be determined from the corresponding behavior of $\Theta_T(z_0)$. We supply some details in a minute. If z_0 is a boundary point, then from the formula for $\Theta_T(\cdot)$ it follows that a necessary condition for z_0 to be in the resolvent set for T is that $\Theta_T(\cdot)$ have a holomorphic extension in a neighborhood of z_0. A further argument shows that this extension must be unitary valued on the circle. Therefore the spectrum of T is the subset S_T of $\bar{\mathbf{D}}$ consisting of the z_0 in $\mathbf{D}$ for which $\Theta_T(z_0)$ fails to be invertible together with the z_0 in $\mathbf{T}$ for which $\Theta_T(\cdot)$ fails to have a holomorphic extension to a neighborhood Ω of z_0 which is unitary on $\Omega \cap \mathbf{T}$. This was first proved in complete generality by Sz.-Nagy and Foiaş **[73]**, but special cases had been obtained earlier by several authors (cf. **[73]**).

THEOREM 5. *If T is a completely nonunitary contraction on $\mathcal{H}$ with characteristic operator function $\Theta_T(\cdot)$ mapping $\mathcal{D}_T$ into $\mathcal{D}_{T^*}$, then $\sigma(T) = S_T$.*

PROOF. In view of the comments made after Proposition 4.1 it is sufficient for the case of interior points to prove this at $z = 0$. If we decompose the domain of T into $\mathcal{D}_T \oplus (\mathcal{H} \ominus \mathcal{D}_T)$ and the range into $\mathcal{D}_{T^*} \oplus (\mathcal{H} \ominus \mathcal{D}_{T^*})$, then corresponding to T there is a 2×2 matrix with operator entries. If the vector f lies in $\mathcal{H} \ominus \mathcal{D}_T$, then $0 = D_T f = (I - T^*T)^{1/2} f$ and hence $T^*Tf = f$ and $\|Tf\| = \|f\|$. Therefore $\|T^*(Tf)\| = \|Tf\|$ which implies that $T(\mathcal{H} \ominus \mathcal{D}_T)$ is contained in $\mathcal{H} \ominus \mathcal{D}_{T^*}$. Since we know already that $T\mathcal{D}_T$ is contained in $\mathcal{D}_{T^*}$, it follows that this matrix is a diagonal matrix with entries mapping $\mathcal{D}_T$ into $\mathcal{D}_{T^*}$ and $\mathcal{H} \ominus \mathcal{D}_T$ onto $\mathcal{H} \ominus \mathcal{D}_{T^*}$ respectively. However, by our previous computation we see that the latter operator is, in fact, an isometric isomorphism of $\mathcal{H} \ominus \mathcal{D}_T$ onto $\mathcal{H} \ominus \mathcal{D}_{T^*}$. Thus T is invertible if and only if $\Theta_T(0)$ is invertible. By our previous remarks this completes the determination of which interior points of $\mathbf{D}$ are in the spectrum of T.

We shall only sketch the proof for the boundary points of $\bar{\mathbf{D}}$. Using the identities $TD_T = D_{T^*} T$ and $T^* D_{T^*} = D_T T^*$ and the formulas for $\Theta_T(\cdot)$ and $\Theta_{T^*}(\cdot)$ given in Proposition 4.1, we obtain the identities

$$\Theta_T(z)D_T = D_{T*}(I - zT^*)^{-1}(z - T)$$

and

$$\Theta_{T*}(w)D_{T*} = D_T(I - wT)^{-1}(w - T^*)$$

for z and w in the resolvent sets of T and T^*, respectively. If both z and $1/\bar{z}$ are in the resolvent set of T, then we obtain

$$\Theta_T(z)\Theta_{T*}(z^{-1})D_{T*} = D_{T*} \quad \text{and} \quad \Theta_{T*}(z^{-1})\Theta_T(z)D_T = D_T$$

which implies that $\Theta_{T*}(z^{-1}) = \Theta_T(z)^{-1}$. If z lies in $\mathbf{T}$ and is in the resolvent set of T, this identity implies that $\Theta_T(\cdot)$ has a holomorphic extension to a neighborhood of z which is unitary on the intersection of this neighborhood with $\mathbf{T}$. The converse is somewhat more difficult to prove. First, one establishes that if $\Theta_T(\cdot)$ has such an extension into a neighborhood beyond $\mathbf{D}$, then the functions $g \oplus h$ in

$$\left\{\left[\mathbf{H}^2_{\mathfrak{D}_T} \oplus \int_{\mathbf{T}} \oplus \overline{\Delta_\lambda L}\frac{d\lambda}{2\pi}\right] \ominus \{\Theta_T f \oplus \Delta f : f \in \mathbf{H}^2\}\right\}$$

have holomorphic extensions into the same neighborhood. Next, one shows that this enables one to extend the resolvent of T holomorphically into the same neighborhood.

The omitted details can be found in [73], which should be consulted in any case for many further aspects of the relation between the spectrum of the canonical model and the characteristic operator function.

Although the contractions in class C_0 are covered by the preceding result, there is a simpler determination of the spectrum which depends only on the minimal function. Before we can discuss this, we need a bit of additional terminology. A scalar inner function φ can be uniquely factored as the product of a Blaschke product B and a singular inner function S (cf. [61]). We define the *support* S_φ of φ to be the union of the closure of the zero set of B and the closed support of the singular measure defining S. An equivalent definition of S_φ is that it is the intersection of $\overline{\mathbf{D}}$ with the complement of the largest open set on which φ exists as a nonvanishing holomorphic function. Since φ has radial limits of modulus one a.e. on $\mathbf{T}$, being holomorphic at a point of $\mathbf{T}$ implies it is unitary. Thus the spectrum of the operator having φ as its characteristic operator function is precisely the support of φ.

PROPOSITION 4.2. *If T is a contraction in class C_0 with minimal function*

m_T, then the spectrum $\sigma(T)$ is equal to S_{m_T}. Moreover, the set $\sigma(T) \cap \mathbf{D}$ consists of eigenvalues of T.

PROOF. If α is a point in $\bar{\mathbf{D}} \backslash S_{m_T}$, then, by our previous remarks, m_T is holomorphic on a neighborhood of α and $m_T(\alpha) \neq 0$. Thus the function defined by

$$u(z) = \frac{m_T(z) - m_T(\alpha)}{z - \alpha}$$

lies in H^∞ and using the functional calculus we obtain

$$u(T)(T - \alpha) = (T - \alpha)u(T) = m_T(T) - m_T(\alpha)I = -m_T(\alpha)I.$$

Therefore, the inverse $(T - \alpha)^{-1}$ exists and equals $-u(T)/m_T(\alpha)$ and hence α is not in the spectrum of T. Thus we have proved that $\sigma(T)$ is contained in S_{m_T}.

Conversely, if α lies in $\mathbf{D}$ and $m_T(\alpha) = 0$, then

$$m_T(z) = \frac{z - \alpha}{1 - \bar{\alpha} z} \varphi(z)$$

for some inner function φ. If $\varphi \equiv 1$, then $T = \alpha$ and hence α is in the spectrum of T. Assume $\varphi \not\equiv 1$. Since $(T - \alpha)\varphi(T)(I - \bar{\alpha}T)^{-1} = m_T(T) = 0$, we obtain $(T - \alpha)f = 0$ for f in the range of $\varphi(T)(I - \bar{\alpha}T)^{-1}$. If this latter subspace consisted of just the zero vector, we would have $\varphi(T) = 0$ contradicting the minimality of m_T. Therefore α is an eigenvalue for T and hence lies in $\sigma(T)$.

Now suppose the intersection E of S_{m_T} and the resolvent set of T were not empty. By the previous paragraph, E would be contained in $\mathbf{T}$. Using the structure theorem for inner functions it is possible to factor m_T as $\varphi_1\varphi_2$ such that φ_2 is not constant and S_{φ_2} is contained in the resolvent set of T. By our previous remarks, this implies that φ_2 is holomorphic and nonvanishing on a neighborhood of the spectrum of T and hence $\varphi_2(T)$ is invertible. Since $\varphi_1(T) = \varphi_2(T)^{-1}m_T(T) = 0$, we again contradict the minimality of m_T, thus completing the proof.

A comparison of the last two results in view of Proposition 3.15 indicates the existence of certain implications concerning holomorphic extensions of matrix inner functions and their determinants. We do not elaborate on this but refer the reader to [**38**] or [**73**].

We conclude our discussion of the spectrum with two corollaries.

COROLLARY 4.3. *If* T *is a contraction in class* C_0 *and* $\{\lambda_i\}_{i\geq 1}$ *are distinct eigenvalues of* T, *then* $\Sigma_{i\geq 1}\ 1 - |\lambda_i| < \infty$.

PROOF. This follows from the fact that the λ_i are the zeros of m_T, together with the corresponding statement concerning the zeros of an inner function (cf. **[61]**).

COROLLARY 4.4. *If* T *is a strict contraction on* H *in class* C_0, *then* T *is algebraic.*

PROOF. Since m_T is an inner function which is holomorphic on a neighborhood of $\overline{\mathbf{D}}$, then m_T is a rational function and the result follows.

This concludes our comments on the relation between the spectrum and the characteristic operator function. Let us point out, however, that much work remains to be done. For example, the fact that H^2 can be represented as a kernel Hilbert space can be viewed as providing a "generalized eigenfunction expansion" for the adjoint of the unilateral shift. Analogous results should be true in the context of canonical models. Further elaboration of this theme occurs in the recent work of Ahern and Clark **[4]**, Clark **[15]**, **[16]** and Kriete **[44]**, **[45]**.

We now consider the role of invariant subspaces in the theory of canonical models. Despite considerable attention, especially during the last several years, the question of the existence of nontrivial invariant subspaces for a general operator on Hilbert space remains unresolved. Much of the original interest in the theory of canonical models stemmed from the belief that it might play the decisive role in solving this problem. Although it would be premature to dismiss this possibility, it seems fair to say that additional ideas will be required to solve the general problem. Nonetheless, two victories can be recorded in the invariant subspace problem for the theory of canonical models: contractions having a "part" in class C_{11} and contractions in class C_0.

Although the original existence proof for contractions having a "part" in class C_{11} depended on a factorization technique due to Lowdenslager, Parrott **[57]** and Sz.-Nagy and Foiaş **[73]** subsequently discovered that a direct elementary proof was possible. We present this proof beginning with the following lemma. Recall that an operator with trivial kernel and dense range is said to be *quasi-invertible.*

LEMMA 4.5. *If* T *is a contraction on* H *in class* $C_{1.}$ *then there exists an isometric operator* V *on a Hilbert space* H_1 *and a quasi-invertible operator* X *from* H *to* H_1 *such that* $XT = VX$.

PROOF. Since $\lim_{n\to\infty}(T^nf, T^nf) = \lim_{n\to\infty}\|T^nf\|^2$ exists for each f in $\mathcal{H}$, it follows using the polarization identity that $\lim_{n\to\infty}(T^nf, T^ng) = \langle f, g\rangle$ exists for f and g in $\mathcal{H}$. Moreover, $\langle\cdot,\cdot\rangle$ defines a positive quadratic form on $\mathcal{H}$ which is definite precisely when T lies in class $C_{1\cdot}$. If we let $\mathcal{H}_1$ denote the completion of $\mathcal{H}$ with respect to this new norm, we obtain a Hilbert space, and setting $Xf = f$ for f in $\mathcal{H}$ defines a contraction from $\mathcal{H}$ into $\mathcal{H}_1$. Moreover, X has dense range and zero kernel and hence is quasi-invertible. If we now define $VXf = XTf$, then

$$\|VXf\|_1 = \|XTf\|_1 = \lim_{n\to\infty}\|T^nTf\| = \lim_{n\to\infty}\|T^nf\| = \|Xf\|_1$$

which implies that V can be extended to an isometry on $\mathcal{H}_1$. Thus the lemma is proved.

We now give the existence proof.

PROPOSITION 4.6. *If T is a contraction on $\mathcal{H}$ for which there exist vectors f and g in $\mathcal{H}$ such that $\lim_{n\to\infty}\|T^nf\| > 0$ and $\lim_{n\to\infty}\|T^{*n}g\| > 0$, then T has a proper invariant subspace.*

PROOF. Since the collection of h in $\mathcal{H}$ for which $\lim_{n\to\infty}\|T^nh\| = 0$ forms an invariant subspace for T smaller than $\mathcal{H}$, we may assume it to be (0) and hence that T lies in class $C_{1\cdot}$. Thus there exists an isometry V_1 on a Hilbert space $\mathcal{H}_1$ and a quasi-invertible operator X from $\mathcal{H}$ into $\mathcal{H}_1$ such that $V_1X = XT$. Since the closure of the range of T is a proper invariant subspace for T, we can assume that the range of T is dense. Thus the range of V_1 is dense and therefore V_1 is unitary.

Analogous consideration of T^* yields a unitary operator V_2 on a Hilbert space $\mathcal{H}_2$ and a quasi-invertible operator Y from $\mathcal{H}$ into $\mathcal{H}_2$ such that $V_2Y = YT^*$. Thus we have the relations $V_1X = XT$ and $Y^*V_2^* = TY^*$, which imply that $V_1XY^* = XTY^* = XY^*V_2^*$. If we write $XY^* = PW$, where P is a positive operator on $\mathcal{H}_1$ and W is an isometric isomorphism from $\mathcal{H}_2$ onto $\mathcal{H}_1$, then we obtain $V_1PW = PWV_2^*$. Multiplying on the right by W, we obtain $V_1P = PWV_2^*W^*$, which implies that $V_1PPV_1^* = PWV_2^*W^*WV_2W^*P = P^2$. Thus we have $V_1P^2 = P^2V_1$, and proceeding as in the beginning of §1, we see that $V_1P = PV_1$. Therefore, we have $PV_1 = V_1P = PWV_2^*W^*$, from which we can conclude that $V_1 = WV_2^*W^*$. Thus the unitary operators are unitraily equivalent and we obtain $V_1X = XT$ and $ZV_1 = TZ$, where $Z = Y^*W^*$.

Now let E be a proper spectral projection for the unitary operator V_1 and let $\mathcal{M}$ denote the span of the ranges of the operators AZE, where A lies in the commutant $(T)'$ of T. If A commutes with T, then so does TA

and hence $\mathcal{M}$ is invariant for T. Moreover, since $XAZV_1 = XATZ = XTAZ = V_1XAZ$, we see that XAZ commutes with E, and hence

$$X\mathcal{M} \subset \bigvee_{A\in(T)'} XAZE\mathcal{H}_1 \subset \bigvee_{A\in(T)'} E(XAZ)\mathcal{H}_1 \subset E\mathcal{H}_1.$$

Thus $\mathcal{M}$ is a proper invariant subspace for T, which completes the proof.

Two operators A and B on Hilbert spaces $\mathcal{A}$ and $\mathcal{B}$ are said to be *quasi-similar* if there exist quasi-invertible operators C and D between $\mathcal{A}$ and $\mathcal{B}$ and between $\mathcal{B}$ and $\mathcal{A}$, respectively, such that $BC = CA$ and $DB = AD$. We have established that T is quasi-similar to the unitary operator V_1. We did not show, however, that if one of a pair of quasi-similar operators has an invariant subspace, then so must the other. Such a result is unknown. A subspace of $\mathcal{H}$ that is invariant for every operator in the commutant of T is said to be *hyperinvariant* for T. If one of two quasi-similar operators has a proper hyperinvariant subspace, then so must the other **[43]**. If the reader checks the preceding proof, he can verify that the subspace $\mathcal{M}$ is actually hyperinvariant for T, and that this subspace was obtained from the subspace $E\mathcal{H}_1$ which is hyperinvariant for V_1 **[25]**.

We now consider invariant subspaces for contraction in class C_0. Suppose T_1 and T_2 are contractions on $\mathcal{H}_1$ and $\mathcal{H}_2$, respectively, in class C_0 with minimal functions m_{T_1} and m_{T_2}. If X is an operator mapping $\mathcal{H}_2$ into $\mathcal{H}_1$ then the matrix $\begin{pmatrix} T_1 & X \\ 0 & T_2 \end{pmatrix}$ defines an operator T on $\mathcal{H}_1 \oplus \mathcal{H}_2$ which under certain circumstances will be a contraction.

LEMMA 4.7. *If T is a contraction, then T belongs to class C_0, and m_T divides the product $m_{T_1}m_{T_2}$.*

PROOF. Observe that $\varphi(T_1) = \varphi(T)|\mathcal{H}_1$ and $\varphi(T_2) = P_{\mathcal{H}_2}\varphi(T)|\mathcal{H}_2$ for φ in H^∞. Thus for h_1 in $\mathcal{H}_1$ we have

$$(m_{T_1}m_{T_2})(T)h_1 = m_{T_2}(T)m_{T_1}(T)h_1 = 0.$$

Similarly, for h_2 in $\mathcal{H}_2$ we have

$$(m_{T_1}m_{T_2})(T)h_2 = m_{T_1}(T)m_{T_2}(T)h_2 = 0$$

since $m_{T_2}(T)h_2$ lies in $\mathcal{H}$. This completes the proof of the lemma.

Our existence result is a corollary to

PROPOSITION 4.8. *If T is a contraction on $\mathcal{H}$ in class C_0 with minimal function m_T, and φ_1 and φ_2 are nonconstant inner functions such that*

$m_T = \varphi_1\varphi_2$, *then* $H_1 = \{h \in H\colon \varphi_1(T)h = 0\}$ *is an invariant subspace for* T *such that the minimal function for* $T|H_1$ *is* φ_1 *and for* $P_{H_2}T|H_2$ *is* φ_2, *where* $H_2 = H \ominus H_1$.

PROOF. First, it is clear that H_1 is a closed subspace of H and is invariant for T, since $\varphi_1(T)Th = T\varphi_1(T)h = 0$ for h in H_1. Secondly, since $\varphi_1(T)\varphi_2(T)H = \varphi(T)H = (0)$, it follows that $\varphi_2(T)H$ is contained in H_1, and thus $P_{H_2}\varphi_2(T)|H_2 = 0$. Therefore,

$$m_{T\,H_1} \text{ divides } \varphi_1, \quad \text{and} \quad m_{P_{H_2}T|H_2} \text{ divides } \varphi_2.$$

From the lemma we see that

$$m_{T|H_1} m_{P_{H_2}T|H_2} \text{ divides } m_T = \varphi_1\varphi_2.$$

Thus φ_1 is a minimal function for $T|H_1$, and φ_2 is a minimal function for $P_{H_2}T|H_2$.

COROLLARY 4.9. *If* T *is a contraction in class* C_0, *then* T *has a proper invariant subspace.*

PROOF. If the minimal function for T can be factored, then the result follows from the proposition. Otherwise, $m_T = (z-\alpha)/(1-\bar{\alpha}z)$ for some α in $\mathbf{D}$, which implies that $T = \alpha I$.

Again the reader can verify that the invariant subspace produced is hyperinvariant except when T is a scalar multiple of the identity.

The techniques just given for producing invariant subspaces seldom yield all of them. There is some reason to believe, however, that it might be possible to generate all the hyperinvariant subspaces for these operators from the constructed ones. This is the case, for example, for finite-dimensional operators.

We have not yet indicated any connection between all the invariant subspaces for a contraction and the characteristic operator function. We consider this problem now. Let us begin with the simpler case of a contraction T in class $C_{.0}$. By Proposition 3.2 it follows that L_* is a complete wandering subspace for W, and hence we can identify $H = \mathbf{H}^2_{L_*} \ominus \Theta_T\mathbf{H}^2_L$. Moreover, the characteristic operator function Θ_T is an isometry from $\mathbf{H}^2_L$ into $\mathbf{H}^2_{L_*}$. If M is an invariant subspace for T, then $M \oplus \mathbf{H}^2_L$ is invariant for W. Further, since M must be pure, there exists a holomorphic unitary function Φ from $\mathbf{H}^2_J$ into $\mathbf{H}^2_{L_*}$ such that $\Phi\mathbf{H}^2_J = M \oplus \mathbf{H}^2_L$, where J is some Hilbert space. Since the range of Θ_T is contained in the range of Φ, we can define $\Psi = \Phi^*\Theta_T$ from $\mathbf{H}^2_L$ into $\mathbf{H}^2_J$. Lastly, Ψ is a holomorphic unitary function satisfying $\Theta_T = \Phi\Psi$. Thus an invariant subspace for T leads to a factoring of Θ_T into the product of two holomorphic unitary functions.

Conversely, if we factor $\Theta_T = \Phi\Psi$, where Φ and Ψ are holomorphic unitary functions from $\mathbf{H}_J^2$ into $\mathbf{H}_{L_*}^2$ and $\mathbf{H}_L^2$ into $\mathbf{H}_J^2$, respectively, then $M = \Phi\mathbf{H}_J^2 \ominus \mathbf{H}_L^2$ is an invariant subspace for T. Moreover, it follows from Corollary 2.4 that $M = (0)$ if and only if Ψ is a constant unitary operator and that $M = H$ if and only if Φ is a constant unitary operator. Since a constant unitary operator is a "unit," one can say that T has no proper invariant subspaces if and only if Θ_T is "prime," where the factors are required to be holomorphic unitary functions. Factoring Θ_T as the product of two nonconstant holomorphic contraction-valued functions does not, in general, lead to a *closed* invariant subspace for T; rather, it leads to a "para-closed" invariant subspace for T, that is, a linear manifold that is the range of some bounded operator (cf. [28]). We do not pursue this notion further.

Let us now consider the general completely nonunitary contraction T on H with invariant subspace M. If we set

$$T_1 = T|M, \qquad T_2 = P_{M^\perp}T|M^\perp \quad \text{and} \quad X = P_M T|M^\perp,$$

then

$$T = \begin{pmatrix} T_1 & X \\ 0 & T_2 \end{pmatrix} \quad \text{on} \quad H = M \oplus M^\perp.$$

Moreover, T_1 and T_2 are completely nonunitary contractions on M and $M^\perp$, respectively.

If W is the minimal unitary dilation for T on $K = G_* \oplus H \oplus G$, then W is a unitary dilation for T_1 on $(G_* \oplus M^\perp) \oplus M \oplus G$ and also is a unitary dilation for T_2 on $G_* \oplus M^\perp \oplus (M \oplus G)$. However, W need not be minimal for either; in particular, one must discard the unitary part of W on $G_* \oplus M^\perp \oplus G$ for T_1 and the unitary part of W on $G_* \oplus M \oplus G$ for T_2. Having done this we obtain the canonical models for T_1 and T_2, respectively, and a context in which to define their respective characteristic operator functions Θ_{T_1} and Θ_{T_2}. We shall not carry out these not entirely routine details but refer the reader to [73, Chapter VII]. We do want to point out, however, that what emerges is that Θ_T is the "product" of Θ_{T_1} and Θ_{T_2}. We have the word product in quotation marks because, more precisely, Θ_T is the product of the holomorphic contractive operator functions defined for T_1 and T_2 relative to the unitary dilation W;[2] and hence these functions are not purely contractive in general, and therefore do not coincide with what we have defined to be the characteristic operator functions for T_1 and T_2.

Now let us consider the reverse situation. Suppose T is a completely nonunitary contraction on H with characteristic operator function $\{\Theta, L, L_*\}$. If we factor $\Theta = \Theta_1\Theta_2$ as the product of two holomorphic contraction-valued functions $\{\Theta_1, J, L_*\}$ and $\{\Theta_2, L, J\}$, then we can construct (cf. [73]) a unitary dilation W for T on a space $K = G_* \oplus M_1 \oplus M_2 \oplus G$, where G and G_* are pure invariant subspaces for W and W^*, respectively, $M_1 \oplus M_2 \oplus G$ and $M_2 \oplus G$ are invariant subspaces for W and H is a subspace of $M_1 \oplus M_2$ such that $H \oplus G$ is invariant for W. Moreover, if

$$T' = P_{M_1 \oplus M_2} W | M_1 \oplus M_2, \qquad T_1 = P_{M_1} W | M_1,$$
$$T_2 = P_{M_2} W | M_2 \qquad \text{and} \qquad X = P_{M_1} W | M_2,$$

then $T' = \begin{pmatrix} T_1 & X \\ 0 & T_2 \end{pmatrix}$ and Θ_1 and Θ_2 are the holomorphic contraction-valued functions defined for T_1 and T_2 relative to the unitary dilation W. The operators T_1 and T_2 are completely nonunitary and the characteristic operator functions Θ_{T_1} and Θ_{T_2} for T_1 and T_2, respectively, are very simply related to the contraction holomorphic functions Θ_1 and Θ_2, respectively; namely, $\Theta_1 = \Theta_{T_1} \oplus V_1$ and $\Theta_2 = \Theta_{T_2} \oplus V_2$ for certain constant isometrical isomorphisms V_1 and V_2. There is now one problem: T' need not be completely nonunitary! The completely nonunitary part of T' coincides with T. Thus if one factors Θ_T, then one obtains, in general, not an invariant subspace for T but an invariant subspace for $T \oplus V$ for some unitary operator V (cf. [73]). (Again we do not supply the details, but it is instructive to consider the case where T_1 is the unilateral shift, T_2 is it adjoint, $T = 0$, and T' is the bilateral shift. The corresponding holomorphic contraction-valued functions are $(0, (0), C)$ for T_1, $(0, C, (0))$ for T_2 and $(0, (0), (0))$ for T.) If somehow one can assert that the operator T_1 has no isometric part[3] or that T_2 has no co-isometric part, then the factorization would have to be *regular*, that is, $H = M_1 \oplus M_2$ and $T = T'$. For example, if one could factor Θ_T as $\Theta_1\Theta_2$

---

[2] Although it is not possible to obtain an arbitrary unitary dilation for T from the canonical model defined by a holomorphic contraction-valued function, it is in this case since the superfluous parts of W for T_1 and T_2 are both bilateral shifts.

[3] This was the essence of the attempt of de Branges and Rovnyak [8] to solve the invariant subspace problem.

where either Θ_1 or Θ_2 is holomorphic in some neighborhood Ω of a point $e^{i\alpha}$ in $\mathbf{T}$ and unitary valued on $\Omega \cap \mathbf{T}$, then such a factorization would necessarily be regular.

Let us summarize the preceding discussion as

THEOREM 6. *If T is a completely nonunitary contraction on $\mathcal{H}$ with characteristic operator function $\{\Theta_T, \mathcal{D}, \mathcal{D}_*\}$, then there is a one-to-one correspondence between the invariant subspaces $\mathcal{M}$ for T and the regular factorizations of Θ_T as the product $\Theta_T = \Phi\Psi$ of two contractive holomorphic operator functions $\{\Phi, \mathcal{E}, \mathcal{D}_*\}$ and $\{\Psi, \mathcal{D}, \mathcal{E}\}$. Moreover, the characteristic operator functions for $T_1 = T|\mathcal{M}$ and $T_2 = P_{\mathcal{M}^\perp}T|\mathcal{M}^\perp$ are the purely contractive parts of Φ and Ψ, respectively.*

The reader should see **[68]** for a different exposition of the relation between factorization and invariant subspaces.

As we have just observed, the invariant subspaces for the contraction T correspond to "certain" factorizations of the characteristic operator function $\{\Theta_T, \mathcal{D}_T, \mathcal{D}_{T^*}\}$ into the product of two contraction-valued holomorphic operator functions. One method of producing an operator with no proper invariant subspace would be to find a "prime" holomorphic contraction-valued operator function. Sz.-Nagy and Foiaş have shown, however, that this is impossible **[73**, p. 214**]**.

A rather different problem which might possess an interesting solution concerns the characterization of those factorizations which lead to hyperinvariant subspaces for T.

Eventually, if one wants to use the invariant subspaces for an operator to decompose it, one is interested in more than the mere existence of invariant subspaces. If the characteristic operator function is scalar-valued, then the standard representation theory for inner and outer functions yields about all one can hope to learn about the structure of the invariant subspaces of the associated contraction. A similar analysis for the matrix case was made by Potapov (cf. **[33]**) and involved a multiplicative integral. (The usual device of converting a multiplicative integral into a ordinary integral via taking logarithms and then exponentiating fails due to the noncommutativity of the matrix values.) To appreciate the degree of subtlety and difficulty in analyzing such factorizations, one should ponder the invariant subspace structure for a finite-dimensional nilpotent operator (cf. **[10]**, **[25]**).

Let us now consider the reducing subspaces for a contraction T on $\mathcal{H}$ in terms of the characteristic operator function. If $\mathcal{M}$ is a reducing subspace for T and we set $T_1 = T|\mathcal{M}$ and $T_2 = T|\mathcal{M}^\perp$, then

$$\mathcal{D}_T = \mathcal{D}_{T_1} \oplus \mathcal{D}_{T_2}, \quad \mathcal{D}_{T^*} = \mathcal{D}_{T_1^*} \oplus \mathcal{D}_{T_2^*} \quad \text{and} \quad \Theta_T = \Theta_{T_1} \oplus \Theta_{T_2}$$

Conversely, if $\mathcal{E}$ and $\mathcal{E}_*$ are subspaces of $\mathcal{D}_T$ and $\mathcal{D}_{T^*}$, respectively, such that $\Theta_T(e^{it})\mathcal{E} \subset \mathcal{E}_*$ and $\Theta_T(e^{it})\mathcal{E}^\perp \subset \mathcal{E}_*^\perp$ a.e., then a reducing subspace is determined for T. We summarize this as

PROPOSITION 4.10. *The reducing subspaces for a contraction T on $\mathcal{H}$ correspond to the common diagonalizations of the collection* $\{\Theta_T(e^{it})\}_{e^{it}\in\mathbf{T}}$.

We leave as an exercise for the reader the task of determining the factorization of Θ_T which corresponds to a reducing subspace.

There is still another way to look at the preceding result. Reducing subspaces correspond to projections in the commutant of the operator. After a certain amount of computing and reflecting, the reader will see that the projection $P_\mathcal{M}$ onto a reducing subspace for T has the form $P_\mathcal{M} = P_\mathcal{H} E | \mathcal{H}$, where E is the projection onto a reducing subspace for the unitary dilation W. Moreover, since $P_\mathcal{M}$ is a projection, it follows that $P_\mathcal{M}^n = P_\mathcal{M} = P_\mathcal{H} E | \mathcal{H} = P_\mathcal{H} E^n | \mathcal{H}$, and hence is a semi-invariant subspace for E. Conversely, it is easy to see that any projection in the commutant W for which $\mathcal{H}$ is a semi-invariant subspace gives rise to a reducing subspace for T. Further, it is clear that the entire selfadjoint algebra that commutes with T is obtained in this manner. However, this phenomena does not end here! The following "lifting theorem" for commutants holds.

THEOREM 7. *If T is a contraction on $\mathcal{H}$ with unitary dilation W on $\mathcal{K}$, then an operator X on $\mathcal{H}$ commutes with T if and only if there exists an operator Y on $\mathcal{K}$ that commutes with W such that $X = P_\mathcal{H} Y | \mathcal{H}$, $\mathcal{H}$ is a semi-invariant subspace for Y, and $\|X\| = \|Y\|$.*

The initial result is this direction is due to Sarason **[60]** and included the case when W is a bilateral shift of multiplicity one. The general result is due to Sz.-Nagy and Foiaş **[73]** (cf. **[24]** for an alternate proof). This result has proved quite useful in the recent work of several authors (**[22]**, **[31]**, **[46]**, **[53]**, **[54]** and **[56]**).

5. Contractions in class C_0 and Jordan models. As we have mentioned in earlier sections, the theory of contractions in class C_0 parallels and subsumes most of the theory of operators on finite-dimensional spaces. In this section we want to sketch some of the recent results of Sz.-Nagy and Foiaş on this class of operators along with some indication of proofs. For a complete picture, the reader should consult the papers cited.

If T is an operator on a finite-dimensional space, then linear algebra provides a Jordan model for T which is similar to T. Now a Jordan model can be decomposed into simpler components in at least two different ways. The more common is to decompose the Jordan model into the direct sum of parts, where each corresponds to a factor in the prime decomposition of the minimal polynomial for T, and consists of the direct sum of the Jordan blocks for that eigenvalue. If T is a contraction in class C_0, whose minimal function m_T is a Blaschke product, then there is an invariant subspace $\mathcal{H}_\lambda$ for T corresponding to each prime factor of m_T. It is, however, not generally possible to conclude that T is similar to the infinite orthogonal direct sum of the parts $T_\lambda = T|\mathcal{H}_\lambda$. The difficulty is the usual one with infinite nonorthogonal direct sums (which is how the $\mathcal{H}_\lambda$ are located in $\mathcal{H}$), and it is unreasonable to expect similarity. The appropriate notion on infinite-dimensional spaces seems to be that of quasi-similarity which was mentioned in the preceding section. Moreover, it is very probable in this case that T is quasi-similar to the infinite orthogonal direct sum of the T_λ.

If the minimal function for T has a singular factor, then no analogue of the preceding discussion is possible since singular inner functions have no prime factorization. One can view the standard representation for inner functions as providing an integral decomposition for singular inner functions in terms of the singular inner functions with one point support. Moreover, it seems reasonable to expect an analogous kind of "analytic direct integral" decomposition for the contraction, perhaps combining the notion of a direct integral with that of a kernel Hilbert space. Further, the factorization results of Potapov should also play a role. All of this is speculation, however, and the progress that has been made in the study of operators in class C_0 has been along different lines. (An exception to this occurs in [27].)

We begin by showing that the notion of quasi-similarity is appropriate for the study of contractions in class C_0.

PROPOSITION 5.1. *If T_1 and T_2 are quasi-similar contractions on $\mathcal{H}_1$ and $\mathcal{H}_2$, respectively, and T_1 lies in class C_0, then so does T_2, and $m_{T_1} = m_{T_2}$.*

PROOF. Since T_1 and T_2 are quasi-similar, there exists a quasi-invertible operator X defined from $\mathcal{H}_1$ to $\mathcal{H}_2$ such that $XT_1 = T_2X$. If $p(z)$ is a polynomial, then $Xp(T_1) = p(T_2)X$. Using the weak *-density of the polynomials in H^∞, the fact that T_1 and T_2 are contractions (Corollary 1.1, and Theorem 4), we see that $X\psi(T_1) = \psi(T_2)X$ for ψ in H^∞. It is now obvious that if T_1 is in class C_0, then T_2 is also. Moreover, we have $m_{T_1} = m_{T_2}$, since X is quasi-invertible.

For normal operators one can prove that quasi-similarity implies unitary equivalence. This is not true for operators in class C_0 and, in fact, Sz.-Nagy and Foiaş have given an example of two quasi-similar operators in class C_0 which are not similar. In view of the previous proposition and Proposition 4.2, it follows that the spectra of two quasi-similar operators in class C_0 coincide. Although quasi-similarity does not preserve the spectrum for general operators, it does for subnormal operators [**17**].

We mentioned that there is a second, less common, decomposition of the Jordan model for an operator T on a finite-dimensional space $\mathcal{H}$ into simpler parts. First, one shows that T is similar to $S_1 \oplus T_2$ acting on $\mathcal{H}_1 \oplus \mathcal{H}_2$, where S_1 possesses a cyclic vector and the minimal polynomial for S_1 coincides with that of T. If one applies this process to T_2 and then iterates this process, one obtains that T is similar to the direct sum $T = \Sigma_{i=1}^n \oplus S_i$, where each S_i possesses a cyclic vector and $m_{S_{i+1}}$ divides m_{S_i} for $i = 1, 2, 3, \cdots, n-1$. It is clear how to obtain the Jordan model for T from this decomposition. What we have done is to decompose T into multiplicity free parts.

Such a decomposition along with the appropriate characterization of "multiplicity free" for operators in class C_0 is the principal achievement of Sz.-Nagy and Foiaş in their study of operators in class C_0. We shall describe these results eventually, but first we introduce some necessary terminology.

If φ is an inner function in H^∞, recall that S_φ denotes the compression of the unilateral shift U_+ to $\mathcal{M}_\varphi = H^2 \ominus \varphi H^2$. A *Jordan model* for operators in class C_0 is defined by inner functions $\varphi_1, \varphi_2, \cdots, \varphi_N$, where φ_{i+1} divides φ_i for $i = 1, 2, \cdots, N-1$ and $1 \leqslant N < \infty$, and is the operator $S_{\varphi_1} \oplus S_{\varphi_2} \oplus \cdots \oplus S_{\varphi_N}$ defined on $\mathcal{M}_{\varphi_1} \oplus \mathcal{M}_{\varphi_2} \oplus \cdots \oplus \mathcal{M}_{\varphi_N}$. Although it is far from obvious, a Jordan model is a quasi-similarity invariant. We shall not prove this but refer the reader to [**72**].

PROPOSITION 5.2. *If*

$$S_{\varphi_1} \oplus S_{\varphi_2} \oplus \cdots \oplus S_{\varphi_N} \quad and \quad S_{\psi_1} \oplus S_{\psi_2} \oplus \cdots \oplus S_{\psi_M}$$

are quasi-similar Jordan models, then $N = M$ *and* $\varphi_i = \psi_i$ *for* $1 \leqslant i \leqslant N$.

A moment's reflection yields that not every contraction in class C_0 is quasi-similar to a Jordan model. For example, if $\mathcal{E}$ is an infinite-dimensional Hilbert space and φ is an inner function, then the compression T of the unilateral shift U_+ to $H^2_{\mathcal{E}} \ominus \varphi H^2_{\mathcal{E}}$ lies in class C_0 with minimal function $m_T = \varphi$, but T is not quasi-similar to a Jordan model. The difficulty, in this case, is clear;

the operator T has "infinite multiplicity," while the multiplicity of a Jordan model is necessarily finite. Although one is tempted to extend the notion of Jordan model to cover this case, the difficulties which arise, in general, from an infinite (nonorthogonal) direct sum are not so easily overcome. It is unknown whether every contraction in class C_0 is quasi-similar to a possibly "infinite" Jordan model.

We avoid these problems by considering only contractions in class C_0 of "finite multiplicity." There are at least two obvious ways to define multiplicity in this context: one in terms of the defect numbers d_T and d_{T^*}, and the other in terms of the number of vectors required to span relative to the operator. In particular, one defines the *multiplicity* μ_T of T on $\mathcal{H}$ as the minimal dimension of a subspace $\mathcal{L}$ of $\mathcal{H}$ for which $\mathcal{H} = \bigvee_{n=0}^{\infty} T^n \mathcal{L}$. (The reader should check the relation of the cardinal μ_N for N a normal operator with the notion of multiplicity introduced in §2.) Although it is not true that $\mu_T = \mu_{T^*}$ for general operators (cf. [36]), it is true for contractions in class C_0.

A relation between the two notions of multiplicity for operators in class $C_{.0}$ is established in

PROPOSITION 5.3. *If T is a contraction on $\mathcal{H}$ in class C_0, then $\mu_T \leqslant d_{T^*}$.*

PROOF. The assumption that T lies in C_0 implies that

$$\lim_{n\to\infty} T^n T^{*n} f = 0$$

for f in $\mathcal{H}$. Thus using the fact that the following series telescopes, we see that $f = \Sigma_{n=0}^{\infty} T^n D_{T^*}^2 T^{*n} f$ for f and $\mathcal{H}$, and hence $\mathcal{H} = \bigvee_{n=0}^{\infty} T^n \mathcal{D}_{T^*}$. Therefore, we obtain $\mu_T \leqslant \dim \mathcal{D}_{T^*} = d_{T^*}$, which concludes the proof.

In general, it is not true that $d_T = \mu_T$ since the former is a unitary invariant, while the latter is a quasi-similarity invariant as the corollary to the following proposition shows.

PROPOSITION 5.4. *If T_1 and T_2 are operators on $\mathcal{H}_1$ and $\mathcal{H}_2$, respectively, and X is a quasi-invertible operator from $\mathcal{H}_1$ to $\mathcal{H}_2$ such that $T_2X = XT_1$, then $\mu_{T_1} \geqslant \mu_{T_2}$.*

PROOF. If $\mathcal{L}$ is a subspace of $\mathcal{H}_1$ having dimension μ_{T_1} such that $\bigvee_{n=0}^{\infty} T_1^n \mathcal{L} = \mathcal{H}_1$, then

$$\mathcal{H}_2 = \operatorname{clos}[X\mathcal{H}_1] = \operatorname{clos}\left[X\left(\bigvee_{n=0}^{\infty} T_1^n \mathcal{L}\right)\right] = \bigvee_{n=0}^{\infty} T_2^n X\mathcal{L}.$$

Therefore, $\mu_{T_2} \leqslant \dim \operatorname{clos}[X\mathcal{L}] = \dim \mathcal{L} = \mu_{T_1}$.

COROLLARY 5.5. *If* T_1 *and* T_2 *are quasi-similar operators, then* $\mu_{T_1} = \mu_{T_2}$.

Now one wants to show that a contraction T in class C_0 possessing a cyclic vector is quasi-similar to S_{m_T}. As the first step in that direction we prove the following.

PROPOSITION 5.6. *If* T *is a contraction on* H *in class* $C_{.0}$, *then there exists a contraction* T_1 *on* H_1 *in class* $C_{.0}$ *and a quasi-invertible operator* X *from* H *to* H_1 *such that* $T_1X = XT$ *and* $d_{T_1^*} = \mu_{T_1} = \mu_T$.

PROOF. Let W on K be the minimal unitary dilation for T. Choose a subspace L in H of dimension μ_T such that $\bigvee_{n=0}^{\infty} T^n L = H$, and set $M = \bigvee_{n=0}^{\infty} W^n L$. Then M is invariant for W, and $W|M$ is a pure isometry. Thus, setting $E = M \ominus WM$, we have $M = \Sigma_{n=0}^{\infty} \oplus W^n E$. Since

$$E = \left(\bigvee_{n=0}^{\infty} W^n L\right) \ominus \left(\bigvee_{n=1}^{\infty} W^n L\right),$$

we see that $\dim E \leqslant \dim L = \mu_T$. Further, since

$$H = \bigvee_{n=0}^{\infty} T^n L = \bigvee_{n=0}^{\infty} P_H W^n L = \operatorname{clos}\,[P_H M] = \bigvee_{n=0}^{\infty} P_H W^n E = \bigvee_{n=0}^{\infty} T^n P_H E,$$

it follows that $\mu_T \leqslant \dim \operatorname{clos}\,[P_H E] \leqslant \dim E$ and therefore $\mu_T = \dim E$.

If we set $M_0 = \{f \in M : P_H f = 0\}$ and $H_1 = M \ominus M_0$, then M_0 is invariant for W since $P_H Wf = TP_H f$ for f in $H \oplus G$, and $H \oplus G$ contains M. Since

$$H = \operatorname{clos}\,[P_H M] = \operatorname{clos}\,[P_H(M_0 \oplus H_1)] = \operatorname{clos}\,[P_H H_1],$$

it follows that $X = P_H|H_1$ is a quasi-invertible operator from H_1 to H. Further, if we set $T_1 = P_{H_1} W|H_1$, then

$$TX = TP_H|H_1 = P_H W|H_1 = P_H P_{H_1} W|H_1 = XT_1.$$

From Propositions 5.2 and 5.3 it follows that $\mu_T \leqslant \mu_{T_1} \leqslant d_{T_1^*}$. Moreover, we have

$$\begin{aligned} T_1 T_1^* &= P_{H_1}(W|M)P_{H_1}(W|M)^*|H_1 \\ &= P_{H_1}(W|M)(W|M)^*|H_1 - P_{H_1}(W|M)(I - P_{H_1})(W|M)^*|H_1 \\ &= P_{H_1}(W|M)(W|M)^*|H_1, \end{aligned}$$

since $P_{H_1}(W|M)P_{M_0}H_1 \subset P_{H_1}WM_0 \subset P_{H_1}M_0 = (0)$, and therefore

$$I - T_1T_1^* = P_{H_1}\{I - (W|M)(W|M)^*\}P_{H_1} = P_{H_1}P_EP_{H_1}.$$

Finally, $d_{T_1^*} = \dim (I - T_1T_1^*)H_1 \leqslant \dim E$, and thus we conclude that

$\mu_T = \mu_{T_1} = d_{T_1^*}$, which completes the proof.

COROLLARY 5.7. *If T is a contraction on H in class C_0, then there exists a contraction T_1 on H_1 in class C_0 and a quasi-invertible operator X from H to H_1 such that $T_1X = XT$ and $d_{T_1} = d_{T_1^*} = \mu_{T_1} = \mu_T$. In particular, if T has a cyclic vector, then T_1 can be taken to be S_{m_T}.*

PROOF. Combine Lemma 5.1 with the proposition.

This is almost enough to establish the quasi-similarity result for cyclic operators in class C_0. What more we need to know is that if such an operator has a cyclic vector, then so does its adjoint. We restrict our attention to the subclass $C_0(N)$ of operators in class C_0 for which $d_T = d_{T^*} = N < \infty$.

If T is a contraction in $C_0(N)$, then one can define the "determinant" δ_T of the characteristic operator function (see the proof of Proposition 3.15). In general, δ_T and the minimal function m_T do not coincide although m_T divides δ_T. The following result of Sz.-Nagy and Foiaş [73], which we shall not prove, provides a simple criterion for when they do.

PROPOSITION 5.8. *If T on H is a contraction in $C_0(N)$, then $\delta_T = m_T$ if and only if T has a cyclic vector.*

COROLLARY 5.9. *If T on H is a contraction in $C_0(N)$, then T has a cyclic vector if and only if T^* has a cyclic vector.*

PROOF. Since $m_{T^*} = \widetilde{m}_T$ and $\delta_{T^*} = \widetilde{\delta}_T$, the result follows.

We can now prove

PROPOSITION 5.10. *If T is a contraction on H in class $C_0(N)$ possessing a cyclic vector, then T is quasi-similar to S_{m_T}.*

PROOF. The result follows by applying Corollary 5.5 to both T and T^*.

This result enables one to classify all cyclic contractions in class $C_0(N)$.

COROLLARY 5.11. *If T_1 and T_2 are cyclic contractions in class $C_0(N)$, then T_1 and T_2 are quasi-similar if and only if $m_{T_1} = m_{T_2}$.*

PROOF. The condition is necessary by Lemma 5.1 and sufficiently follows from the preceding proposition.

Actually, Sz.-Nagy and Foiaş have obtained [**71**] several properties equivalent to an operator being multiplicity free. We do not provide a proof but we state this as

THEOREM 8. *If* T *is a contraction on* $\mathcal{H}$ *in class* $C_0(N)$, *then the following properties are equivalent*:

(i) T *has a cyclic vector.*

(i*) T^* *has a cyclic vector.*

(ii) T *is quasi-similar to* S_{m_T}.

(iii) *For each inner divisor* m' *of* m_T, *there exists a unique invariant subspace* $\mathcal{L}$ *for* T *such that* $m_{T|\mathcal{L}} = m'$, *namely the hyperinvariant subspace* $\mathcal{L} = \{f \in \mathcal{H}: m'(T)f = 0\}$.

(iv) *There does not exist a proper invariant subspace* $\mathcal{L}$ *for* T *such that* $m_{T|\mathcal{L}} = m_T$.

(v) *There do not exist two different invariant subspaces* $\mathcal{L}_1$ *and* $\mathcal{L}_2$ *for* T *such that* $T|\mathcal{L}_1$ *and* $T|\mathcal{L}_2$ *are quasi-similar.*

(vi) *The commutant of* T *is commutative.*

In particular, the commutant for T is shown to consist of functions in T, where the functions lie in a certain algebra of quotients of bounded holomorphic functions on **D**.

The extension of this model to operators which are not multiplicity free depends on the following.

PROPOSITION 5.12. *If* T *is a contraction on* $\mathcal{H}$ *of finite multiplicity then there exists a cyclic invariant subspace* $\mathcal{H}_0$ *for* T *such that the minimal function* $m_{T|\mathcal{H}_0} = m_T$.

PROOF. Since there exists a finite collection of vectors, $f_1, f_2, \cdots, f_k$ such that

$$\bigvee_{1 \leqslant i \leqslant k;\, 0 \leqslant n < \infty} T^n f_i = \mathcal{H},$$

it is sufficient to show that every subspace of $\mathcal{H}$, invariant for T and generated by two vectors, is cyclic.

Suppose $\mathcal{H}_0$ is an invariant subspace for T with cyclic vector f in $\mathcal{H}_0$ such that $m_{T|\mathcal{H}_0} = \varphi_0$. If $\varphi_0 = m_T$, then we are finished; assume therefore that $\varphi_0 \neq m_T$ and that g is a vector in $\mathcal{H}$ such that $m_{T|\mathcal{H}_\infty} = \psi \neq \varphi_0$, where $\mathcal{H}_\infty = \mathcal{H}_0 \vee (\bigvee_{n=0}^{\infty} T^n g)$. For α in **C** set $h_\alpha = f + \alpha g$ and let φ_α be the minimal function for $T|\mathcal{H}_\alpha$, where $\mathcal{H}_\alpha = \bigvee_{n=0}^{\infty} T^n (f + \alpha g)$.

Since $\mathcal{H}_\alpha \vee \mathcal{H}_\beta = \mathcal{H}_\infty$ for $\alpha \neq \beta$, it follows that the least common multiple of φ_α and φ_β is ψ. This implies that $\theta_\alpha = \psi\bar{\varphi}_\alpha$ and $\theta_\beta = \psi\bar{\varphi}_\beta$ are relatively prime for $\alpha \neq \beta$. Since the number of relatively prime inner divisors of ψ is countable (use the representation for inner functions), it follows that $\varphi_2 = \psi$ for all but countably many α, which concludes the proof.

This proof is based on arguments due to Herrero **[41]** and Sherman **[66]** The result is due independently to Sz.-Nagy and Foiaş **[71]** and Herrero **[41]**.

The preceding result enables one to split off a maximal multiplicity free invariant subspace. If the operator has finite multiplicity, then this process must terminate and we are led to a Jordan model. This is a vast oversimplification of what is involved and the reader should consult [72] for details. We state the final result as

THEOREM 9. *If T is a contraction on $\mathcal{H}$ in class C_0 and of finite multiplicity μ_T, then there exists a Jordan model $S_{\varphi_1} \oplus S_{\varphi_2} \oplus \cdots \oplus S_{\varphi_N}$ on $\mathcal{M}_{\varphi_1} \oplus \mathcal{M}_{\varphi_2} \oplus \cdots \oplus \mathcal{M}_{\varphi_N}$ quasi-similar to T such that $\varphi_1 = m_T$ and $N = \mu_T$. Moreover, two such contractions are quasi-similar if and only if these Jordan models coincide.*

An alternate derivation of this result paralleling methods from linear algebra more closely has been obtained by Moore and Nordgren **[52]**, **[55]**.

Most of the results in this section requiring the hypothesis that the operators lie in $C_0(N)$ can be extended to general operators in C_0 **[74]**. The key to this extension requires an argument which forms the key to the solution of Sz.-Nagy and Foiaş **[75]** of a conjecture of Sherman **[65]**.

PROPOSITION 5.13. *If T is a contraction on $\mathcal{H}$ such that the restriction of T to every cyclic invariant subspace is in C_0, then T is in C_0.*

Sherman's original conjecture took a different form and concerned operator-valued inner functions. An extension of this result has been given by Sherman **[66]** and a different proof has been given by Helson **[39]**.

Although we have not been able to describe all the results known about operators in class C_0, we have given the main ones. It is clear that this is a well-behaved class of operators which we are well on the way toward understanding.

6. Related and more general models. Although we have concentrated exclusively on the canonical model of Sz.-Nagy and Foiaş, related results were obtained concurrently by many authors. In some cases it is only recently that the relation has been recognized.

The original paper on canonical models is due to Livšic **[50]**, who was then joined by Brodskiĭ **[12]**. They considered (possibly unbounded) operators with "small" imaginary part, that is, operators that are nearly selfadjoint. The relation of this work to that of Sz.-Nagy and Foiaş is established using the Cayley transform, and our assumption that the operator is a contraction corresponds to the assumption that the operator is dissipative, that is, that the operator has nonnegative imaginary part. This work has been taken up by a number of Russian authors and is presented in the recent monographs of Brodskiĭ **[11]**, and Gohberg and Kreĭn **[32]**, **[33]**. A model theory based on J-unitary dilations for operators which are not contractions has been begun by Davis **[18]** and Davis and Foiaş **[19]**. This corresponds to the nearly selfadjoint operators with nondefinite imaginary part.

A theory of canonical models closely related to that of Sz.-Nagy and Foiaş was formulated by de Branges and Rovnyak **[8]**, **[9]**. Their model assigns a contraction to a holomorphic contraction-valued function on **D**, and coincides with that of Sz.-Nagy and Foiaş for the completely nonisometric case but not otherwise (cf. **[29]**). Formal power series with vector-valued and operator-valued coefficients are basic to their approach.

During the past decade and a half, Lax and Phillips **[49]** have developed an approach to the study of the scattering of waves by obstacles. Although this phenomenon is described by certain systems of hyperbolic partial differential equations, the mathematical framework in which these equations are studied is essentially equivalent to that of the theory of canonical models. The problems emphasized by the two theories are, however, usually quite different.

A further occurence of this theory has been in engineering systems theory. Associated with every system is an operator and the properties of the system can be derived from the corresponding characteristic operator function. The assumption that the operator is a contraction corresponds to assuming that the system does not produce energy. The basic problem is realizing or synthesizing a system with given properties; mathematically this corresponds to constructing an operator from its characteristic operator function and in this context was solved by Ho and Kelman (cf. **[14]**). An earlier development of a connection between systems theory and the theory of canonical models was given by Livšic in **[51]**.

Lastly, a categorical approach to model theory was outlined by Foiaş in **[29]**, while constructive approaches to the theory of canonical models have been begun by Ahern and Clark **[4]**, Clark **[15]**, **[16]** and Kriete **[44]**, **[45]**. The homomorphisms constructed by these latter authors are "generalized Fourier transforms."

All of the preceding model theories are closely related and can be viewed as being based on the unit disk or half plane. Moreover, the universality of these models can be deceiving since the theory is really successful only for the class of "nearly unitary" operators and the operators of class C_0. Thus it seems reasonable, and perhaps even necessary, to consider models based on more complicated subsets of the complex plane.

A fundamental step in this direction was made by Sarason **[59]** who showed that the Sz.-Nagy-Foiaş theory could be extended to operators having a subset X of $\mathbf{C}$ as a spectral set, where the interior of X has a connected complement. Such an X decomposes into the countable union of connected, simply connected domains and the operator has a corresponding direct sum decomposition. Sarason shows, using conformal mapping and the extended functional calculus, how to reduce the study of such operators to that of contractions. Thus, what is important is the conformal type of a region.

The extension of model theory to regions of higher connectivity has scarcely been started. First of all, the problem of determining the class of which operators would have such a model is unsolved. Clearly, such an operator would have X as a spectral set, but the converse is unknown. What one needs to show is that an operator T having X as a spectral set has a normal power dilation N such that the spectrum of N is contained in the boundary of X. The corresponding result for X having a connected complement was established by Berger, Foiaş and Lebow (cf. **[73]**). Arveson has shown **[5]** that such operators are characterized by having X as a "complete spectral set."

Ignoring this problem and continuing, if one wants to follow the program given in our exposition, then one needs to know rather completely the invariant subspace structure of normal operators having their spectrum contained in the boundary of X. Some results in this direction are known **[1]**, **[2]**, **[37]** and others will appear in **[3]**. It is not yet clear where all of this will lead, but it is closely connected with certain other topics such as operator representations of function algebras.

BIBLIOGRAPHY

1. M. B. Abrahamse, *Toeplitz operators in multiply connected domains,* Bull. Amer. Math. Soc. **77** (1971), 449-454. MR **42** #8313.

2. ———, *Toeplitz operators in multiply connected domains,* Thesis, University of Michigan, Ann Arbor, Mich., 1971.

3. M. B. Abrahamse and R. G. Douglas, *Subnormal operators related to multiply connect domains,* Advances in Math. (to appear).

4.. P. R. Ahern and D. N. Clark, *On functions orthogonal to invariant subspaces,* Acta. Math. **124** (1970), 191-204. MR **41** #8981a.

5. W. B. Arveson, *Subalgebras of C^*-algebras.* I, II, Acta Math. **123** (1969), 141–224; ibid. **128** (1972), 271–308. MR **40** #6274.

6. ———, *Representations of C^*-algebras* (to appear).

7. A. Beurling, *On two problems concerning linear transformations in Hilbert space,* Acta Math. **81** (1949), 239–255.

8. L. de Branges and J. Rovnyak, *The existence of invariant subspaces,* Bull. Amer. Math. Soc. **70** (1964), 718–721; ibid **71** (1965), 396. MR **29** #6279.

9. ———, *Canonical models in quantum scattering theory,* Perturbation Theory and its Applications in Quantum Mechanics (Proc. Adv. Sem. Math. Res. Center, U. S. Army, Theoret. Chem. Inst., Univ. of Wisconsin, Madison, Wis., 1965), Wiley, New York, 1966, pp. 295–392. MR **39** #6109.

10. L. Brickman and P. Fillmore, *The invariant subspace lattice of a linear transformation,* Canad. J. Math. **19** (1967), 810-822. MR **35** #4242.

11. M. S. Brodskiĭ, *Triangular and Jordan representations of linear operators,* "Nauka", Moscow, 1969; English transl., Transl. Math. Monographs, vol. 32, Amer. Math. Soc., Providence, R. I., 1971. MR **41** #4283.

12. M. S. Brodskiĭ and M. S. Livšic, *Spectral analysis of non-selfadjoint operators and intermediate systems,* Uspehi Mat. Nauk **13** (1958), no. 1 (79), 3–84; English transl., Amer. Math. Soc. Transl. (2) **13** (1960), 265–346. MR **20** #7221; **22** #3982.

13. N. G. deBruijn, *On unitary equivalence of unitary dilations of contractions on Hilbert space,* Acta Sci. Math. (Szeged) **23** (1962), 100–105. MR **26** #6794.

14. C. T. Chen, *Introduction to linear systems theory,* Holt, Rinehart, and Winston, New York, 1970.

15. D. N. Clark, *Extending Fourier transforms into Sz.-Nagy–Foiaş spaces,* Bull. Amer. Math. Soc. **78** (1972), 65–67. MR **44** #7244.

16. ———, *Concrete model theory for a class of operators,* J. Functional Analysis **14** (1973), 269–280.

17. S. Clary, *On quasi-similarity of subnormal operators,* Thesis, University of Michigan, Ann Arbor, Mich., 1973.

18. C. Davis, *J-unitary dilation of a general operator,* Acta Sci. Math. (Szeged) **31** (1970), 75–86. MR **41** #9032.

19. C. Davis and C. Foiaş, *Operators with bounded characteristic functions and their J-unitary dilations,* Acta Sci. Math. (Szeged) **32** (1971), 127–140.

20. R. G. Douglas, *Structure theory for operators.* I, J. Reine Angew. Math. **232** (1968), 180–193. MR **38** #6390.

21. ———, *Notes on multiplicity theory,* Lecture Notes, University of Michigan, Ann Arbor, Mich., 1969.

22. ———, *On the operator equation* $S^*XT = X$ *and related topics,* Acta Sci. Math. (Szeged) **30** (1969), 19–32. MR **40** #3347.

23. ———, *Banach algebra techniques in operator theory,* Pure and Appl. Math., Academic Press, New York, 1972.

24. R. G. Douglas, P. Muhly and C. M. Pearcy, *Lifting commuting operators,* Michigan Math. J. **15** (1968), 385–395. MR **38** #5046.

25. R. G. Douglas and C. M. Pearcy, *On a topology for invariant subspaces,* J. Functional Analysis **2** (1968), 323–341. MR **38** #1547.

26. P. Duren, *Theory of* H^p *spaces,* Academic Press, New York, 1970. MR **42** #3552.

27. C. Foiaş, *The class* C_0 *in the theory of decomposable operators,* Rev. Romain Math. Pures Appl. **14** (1969), 1433–1440.

28. ———, *Invariant para-closed subspaces,* Indiana Univ. Math. J. **21** (1971/72), 887–906. MR **45** #2516.

29. ———, *Some applications of structural models for operators on Hilbert space,* Proc. Internat. Congress Math. (Nice, 1970), vol. 2, Gauthier–Villars, Paris, 1971, pp. 433–440.

30. C. Foiaş and I Suciu, *Szegö measures and spectral theory in Hilbert spaces,* Rev. Roumaine Math. Pures Appl. **11** (1966), 147–159. MR **34** #3363.

31. C. Foiaş and B. Sz.-Nagy, *The "Lifting Theorem" for intertwining operators and some new applications,* Indiana Univ. Math. J. **20** (1971), 901–904.

32. I. C. Gohberg and M. G. Kreĭn, *Introduction to the theory of linear nonselfadjoint operators in Hilbert space,* "Nauka", Moscow, 1965; English transl., Transl. Math. Monographs, vol. 18, Amer. Math. Soc., Providence, R. I., 1967. MR **36** #3137; **39** #7447.

33. ———, *Theory and applications of Volterra operators in Hilbert space,* "Nauka", Moscow, 1967; English transl., Transl. Math. Monographs, vol. **24**, Amer. Math. Soc., Providence, R. I., 1970. MR **36** #2007.

34. P. R. Halmos, *Normal dilations and extensions of operators,* Summa Brasil. Math. **2** (1950), 125–134. MR **13**, 359.

35. ———, *Shifts on Hilbert spaces,* J. Reine Angew. Math. **208** (1961), 102–112. MR **27** #2868.

36. ———, *A Hilbert space problem book,* Van Nostrand, Princeton, N. J., 1967. MR **34** #8178.

37. M. Hasumi, *Invariant subspace theorems for finite Riemann surfaces,* Canad. J. Math. **18** (1966), 240–255. MR **32** #8200.

38. H. Helson, *Lectures on invariant subspaces,* Academic Press, New York, 1964. MR **30** #1409.

39. ———, *Boundedness from measure theory,* Linear Operators and Approximation (P. L. Butzer, J. P. Kahane and B. Sz.-Nagy, Ed.), Birkhauser Verlag, Basel and Stuttgart, 1972.

40. H. Helson and D. Lowdenslager, *Prediction theory and Fourier series in several variables,* Acta Math. **99** (1958), 165–202. MR **20** #4155.

41. D. A. Herrero, *Inner function-operators,* Dissertation, University of Chicago, Chicago, Ill., 1970.

42. K. Hoffman, *Banach spaces of analytic functions,* Prentice-Hall Ser. in Modern Analysis, Prentice-Hall, Englewood Cliffs, N. J., 1962. MR **24** #A2844.

43. T. Hoover, *Hyperinvariant subspaces for n-normal operators,* Acta Sci. Math. (Szeged) **32** (1971), 109–120.

44. T. L. Kriete, III, *A generalized Paley-Wiener theorem,* J. Math. Anal. Appl. **36** (1971), 529–555. MR **44** #5473.

45. ———, *Complete non-self adjointness of almost self adjoint operators,* Pacific J. Math. **45** (1972), 413–437.

46. T. L. Kriete, III, B. Moore, III and B. Page, *Compact intertwining operators,* Michigan Math. J. **18** (1971), 115–119. MR **44** #822.

47. P. Lax, *Translation invariant spaces,* Acta Math. **101** (1959), 163–178. MR **21** #4359.

48. ———, *Translation invariant spaces,* Proc. Internat. Sympos. Linear Spaces (Jerusalem, 1960), Jerusalem Academic Press, Jerusalem; Pergamon, Oxford, 1961, pp. 299–306. MR **25** #4345.

49. P. Lax and R. S. Phillips, *Scattering theory,* Pure and Appl. Math., vol. 26, Academic Press, New York, 1967. MR **36** #530.

50. M. S. Livšic, *On a certain class of linear operators in a Hilbert space,* Math. Sb. **19** (61) (1946), 239–262; English transl., Amer. Math. Soc. Transl. (2) **13** (1960), 61–83. MR **8**, 588; **22** #3981a.

51. ———, *Operators, oscillations, waves. Open systems,* "Nauka", Moscow, 1966; English transl., Transl. Math Monographs, vol. 34, Amer. Math. Soc., Providence, R. I., 1973. MR **38** #1922.

52. B. Moore, III and E. A. Nordgren, *On quasi-equivalence and quasi-similarity,* Acta Sci. Math. (Szeged) **34** (1973), 311–316.

53. B. Moore, III and L. B. Page, *The class Ω of operator weighted functions,* J. Math. Mech. **19** (1970), 1011–1017. MR **43** #2545.

54. P. S. Muhly, *Compact operators in the commutant of a contraction,* J. Functional Analysis **8** (1971), 197–224.

55. E. A. Nordgren, *On quasi-equivalence of matrices over H^∞,* Acta Sci. Math. (Szeged) (to appear).

56. L. B. Page, *Applications of the Sz.-Nagy and Foiaş lifting theorem,* Indiana Univ. Math. J. **20** (1970), 135–145. MR **41** #6003.

57. S. Parrott, *On an invariant subspace theorem,* 1965. (preprint)

58. ———, *Unitary dilations for commuting contractions,* Pacific J. Math. **34** (1970), 481–490. MR **42** #3607.

59. D. E. Sarason, *On spectral sets having connected complement,* Acta Sci. Math. (Szeged) **26** (1965), 289–299. MR **32** #6229.

60. ———, *Generalized interpolation in* H^∞, Trans. Amer. Math. Soc. **127** (1967), 179–203. MR **34** #8193.

61. ———, *Invariant subspaces,* this volume, pp. 1–45.

62. J. J. Schäffer, *On unitary dilations of contractions,* Proc. Amer. Math. Soc. **6** (1955), 322. MR **16**, 934.

63. M. Schreiber, *Unitary dilations of operators,* Duke Math. J. **23** (1956), 579–594. MR **18**, 748.

64. ———, *A functional calculus for general operators in Hilbert space,* Trans. Amer. Math. Soc. **87** (1958), 108–118. MR **20** #6040.

65. M. J. Sherman, *Invariant subspaces containing all analytic directions,* J. Functional Analysis **3** (1969), 164–172.

66. ———, *Invariant subspaces containing all constant directions,* J. Functional Analysis **8** (1971), 82–85. MR **45** #951.

67. B. Sz.-Nagy, *Sur les contractions de l'espace de Hilbert,* Acta Sci. Math. (Szeged) **15** (1963), 87–92. MR **15**, 326.

68. ———, *Sous-espaces invariants d'un opérateur et factorisations des à fonction caractéristique,* Proc. Internat. Congress Math. (Nice 1970), vol. 2, Gauthier-Villars, Paris, 1971, pp. 459–465.

69. B. Sz.-Nagy and C. Foiaş, *Sur les contractions de l'espace de Hilbert.* IV, Acta Sci. Math. (Szeged) **21** (1960), 251–259. MR **23** #A3445.

70. ———, *Vecteurs cycliques et quasi-affinités,* Studia Math. **31** (1968), 35–42. MR **38** #5050.

71. ———, *Opérateurs sans multiplicité,* Acta Sci. Math. (Szeged) **30** (1969), 1–18.

72. ———, *Modèle de Jordan pour une classe d'opérateurs de l'espace de Hilbert,* Acta Sci. Math. (Szeged) **31** (1970), 91–115. MR **41** #9038.

73. ———, *Analyse harmonic des opèrateurs de l'espace de Hilbert,* Masson, Paris; Akad. Kiadó, Budapest, 1967; English rev. transl., North–Holland, Amsterdam; American Elsevier, New York; Akad. Kiadó, Budapest, 1970. MR **37** #778; **43** #947.

74. ———, *Compléments à l'étude des opérateurs de class* C_0, Acta. Sci. Math. (Szeged) **31** (1970), 287–296. MR **44** #845.

75. ———, *Local characterization of operators of class* C_0, J. Functional Analysis **8** (1971), 76–81. MR **45** #950.

STATE UNIVERSITY OF NEW YORK AT STONY BROOK

Mathematical Surveys
Volume 13
1974

V

A SURVEY OF THE LOMONOSOV TECHNIQUE IN THE THEORY OF INVARIANT SUBSPACES

BY

CARL PEARCY AND ALLEN L. SHIELDS

AMS (MOS) subject classifications (1970). Primary 47A02; Secondary 47A15.

Let X be an infinite-dimensional, complex Banach space, and let $L(X)$ denote the algebra of all bounded linear operators on X. A *subspace* of X, is, by definition, a closed linear manifold in X. If $A \in L(X)$, then a subspace M of X is said to be a *nontrivial invariant subspace* for A if $(0) \neq M \neq X$ and $AM \subset M$. Likewise, a subspace M of X is a *nontrivial hyperinvariant subspace* for A if $(0) \neq M \neq X$ and $A'M \subset M$ for every operator A' in $L(X)$ that commutes with A. The question of whether every operator [every nonscalar operator] in $L(X)$ has a nontrivial invariant [hyperinvariant] subspace is an open problem, called the *invariant [hyperinvariant] subspace problem.* Despite the fact that these questions are starkly simple to state, they have proved intractable for several decades. (A good account of the status of these problems prior to 1973 can be found in the forthcoming book on the subject by Radjavi and Rosenthal [1].)

Recently, the young mathematician Victor Lomonosov introduced an elegant technique [8] which enabled him to solve some hard problems in the theory of invariant subspaces that pertain to compact operators. (An operator K in $L(X)$ is *compact* if K maps the closed unit ball in X onto a set in X that is precompact in the norm topology.)

It is the purpose of this expository note to make Lomonosov's contribution available in English, and at the same time to point out some further consequences of his method of proof. We have incorporated the ideas of a number of colleagues: W. Averson, R. G. Douglas, M. Hilden, and T. Ito. If $\mathcal{A}$ is any subalgebra of $L(X)$, we denote by $\mathcal{A}'$ the commutant of $\mathcal{A}$ (i.e., $\mathcal{A}' = \{A' \in L(X): A'A = AA' \text{ for every } A \text{ in } \mathcal{A}\}$), and we write $\mathcal{A}''$ for $(\mathcal{A}')'$. It is elementary that if $\mathcal{A}$ is any subalgebra of $L(X)$, then $\mathcal{A} \subset \mathcal{A}''$ and $\mathcal{A}' = \mathcal{A}'''$. Furthermore, if $\mathcal{A}$ is abelian, then $\mathcal{A} \subset \mathcal{A}'' \subset \mathcal{A}'$, and $\mathcal{A}''$ is also abelian. A subspace M of X is said to be *invariant* for a subalgebra $\mathcal{A}$ of $L(X)$ if $AM \subset M$ for every A in $\mathcal{A}$, and $\mathcal{A}$ is said to be *transitive* if the only subspaces invariant for $\mathcal{A}$ are (0) and X.

We begin this expository account with a fixed point result for nonlinear mappings that is basic to what follows. The use of this result was one of the brilliant new ideas of Lomonosov.

PROPOSITION 1. *Let* Φ *be a function defined and (norm) continuous on a closed convex set* $\mathcal{C}$ *in the Banach space* $\mathcal{X}$, *and suppose that* $\Phi(\mathcal{C})$ *is contained in a (norm) compact subset* $\mathcal{K}$ *of* $\mathcal{C}$. *Then there exists a point* x_0 *in* $\mathcal{C}$ *such that* $\Phi(x_0) = x_0$.

PROOF. By virtue of Mazur's theorem [9, p. 416], the closed convex hull $\overline{\text{co}}(\mathcal{K})$ is a compact, convex subset of $\mathcal{X}$, and since $\mathcal{C}$ is convex and closed, clearly $\overline{\text{co}}(\mathcal{K}) \subset \mathcal{C}$. Since $\Phi(\mathcal{C}) \subset \mathcal{K}$, we have $\Phi(\overline{\text{co}}(\mathcal{K})) \subset \mathcal{K} \subset \overline{\text{co}}(\mathcal{K})$, and the result now follows from the Schauder-Tychonoff fixed point theorem [3, p. 456].

The following theorem may be regarded as the centerpiece of Lomonosov's elegant new technique. Although it is not stated in [8], its proof is all there.

THEOREM 2. *If* A *is a transitive subalgebra of* $\mathcal{L}(\mathcal{X})$ *and if* K *is any nonzero compact operator in* $\mathcal{L}(\mathcal{X})$, *then there exists an operator* A *in* A *such that the operator* AK *has* 1 *as an eigenvalue (i.e., there exists a nonzero vector* x *in* $\mathcal{X}$ *such that* $AKx = x$). *Likewise, there exists an operator* B *in* A *such that* KB *has* 1 *as an eigenvalue.*

PROOF. Without loss of generality we may assume that $\|K\| = 1$. Let x_0 be any vector in $\mathcal{X}$ such that $\|Kx_0\| > 1$, and observe that this implies that $\|x_0\| > 1$. Let $\mathcal{S}$ be the closed unit ball in $\mathcal{X}$ with center x_0 (i.e., $\mathcal{S} = \{x \in \mathcal{X}: \|x - x_0\| \leqslant 1\}$), and let $\mathcal{D}$ denote the norm closure of the set $K(\mathcal{S})$. Then $\mathcal{D}$ is compact and convex, and it is clear that $0 \notin \mathcal{S}$ and $0 \notin \mathcal{D}$. We define the open sets

$$\mathcal{U}_T = \{y \in \mathcal{X}: \|Ty - x_0\| < 1\}, \qquad T \in \mathcal{L}(\mathcal{X}),$$

and we assert that $\bigcup_{T \in \mathsf{A}} \mathcal{U}_T = \mathcal{X} \backslash (0)$. This can be seen as follows. For any nonzero y in $\mathcal{X}$, the linear manifold $\mathsf{A}y$ is invariant for A, and thus its closure $[\mathsf{A}y]^-$ satisfies $[\mathsf{A}y]^- = \mathcal{X}$, since A is transitive. Hence there is some T in A such that $\|Ty - x_0\| < 1$. In particular, then, we have

$$\mathcal{D} \subset \mathcal{X} \backslash (0) = \bigcup_{T \in \mathsf{A}} \mathcal{U}_T,$$

and since $\mathcal{D}$ is compact, there exist $T_1, \cdots, T_n$ in A such that

$$\mathcal{D} \subset \mathcal{U}_{T_1} \cup \cdots \cup \mathcal{U}_{T_n}. \tag{1}$$

We now construct a partition of unity for this open covering of $\mathcal{D}$ by defining

$$\alpha_j(y) = \max\,[0, 1 - \|T_j y - x_0\|], \qquad y \in \mathcal{D}, \quad j = 1, \cdots, n.$$

Note that the α_j are continuous on $\mathcal{D}$, that $0 \leqslant \alpha_j \leqslant 1$, and that $\Sigma_{j=1}^n \alpha_j(y) > 0$ for every y in $\mathcal{D}$. Thus we may and do define

$$\beta_j(y) = \alpha_j(y) \Big/ \sum_{i=1}^n \alpha_i(y), \qquad y \in \mathcal{D}, \quad j = 1, \cdots, n,$$

and it is immediate that the β_j are continuous on $\mathcal{D}$, that $0 \leqslant \beta_j \leqslant 1$, and that $\Sigma_{j=1}^n \beta_j(y) \equiv 1$ on $\mathcal{D}$. The function Ψ mapping $\mathcal{D}$ into $\mathfrak{X}$ is now defined by

$$\Psi(y) = \sum_{j=1}^n \beta_j(y) T_j y, \qquad y \in \mathcal{D},$$

and it is routine to verify that Ψ is continuous on $\mathcal{D}$. Furthermore, since for every x in $\mathcal{S}$ we may write $x_0 = \Sigma_{j=1}^n \beta_j(Kx)x_0$, we have

$$\|\Psi(Kx) - x_0\| \leqslant \sum_{j=1}^n |\beta_j(Kx)|\, \|T_j Kx - x_0\| \leqslant 1,$$

from which it follows that $\Psi(K\mathcal{S}) \subset \mathcal{S}$, and since Ψ is continuous and $\mathcal{S}$ is closed, $\Psi(\mathcal{D}) \subset \mathcal{S}$. Thus, if we define Φ_1 on $\mathcal{S}$ to be the function $\Phi_1(y) = \Psi(Ky)$, $y \in \mathcal{S}$, then Φ_1 maps $\mathcal{S}$ into itself, and it is easy to check that Φ_1 is continuous and that the range of Φ_1 is a precompact subset of $\mathcal{S}$. Hence, by Proposition 1 there exists a (necessarily nonzero) vector y_1 in $\mathcal{S}$ such that $\Phi_1(y_1) = y_1$, and if we define $A = \Sigma_{j=1}^n \beta_j(Ky_1)T_j$, then clearly $A \in \mathfrak{A}$ and $AKy_1 = y_1$ as desired.

To establish the second conclusion of the theorem, define Φ_2 on $\mathcal{D}$ by $\Phi_2(y) = K(\Psi(y))$, $y \in \mathcal{D}$. Then Φ_2 is continuous and maps $\mathcal{D}$ into itself. It follows from Proposition 1 (or directly from the Schauder-Tychonoff theorem) that there exists a (necessarily nonzero) vector y_2 in $\mathcal{D}$ such that $\Phi_2(y_2) = y_2$. Thus, if we define $B = \Sigma_{j=1}^n \beta_j(y_2)T_j$, then $B \in \mathfrak{A}$ and $KBy_2 = y_2$. This completes the proof of the theorem.

An immediate corollary of Theorem 1 is the following marvelous result, whose statement appears in [8] as an "added in proof."

THEOREM 3. *Let* A *be any nonscalar operator in* $\mathcal{L}(\mathfrak{X})$, *and suppose that* A *commutes with a nonzero compact operator* K. *Then* A *has a nontrivial hyperinvariant subspace.*

PROOF. Let $\mathcal{A}$ denote the algebra consisting of all polynomials $p(A)$. What must be shown is that the algebra $\mathcal{A}'$ is not transitive, since $\mathcal{A}'$ is exactly the commutant of A. We suppose, to the contrary, that $\mathcal{A}'$ is transitive. By applying Theorem 1 to $\mathcal{A}'$ and K we conclude that there exists an operator A' in $\mathcal{A}'$ such that 1 is an eigenvalue of the compact operator $A'K$, with associated finite-dimensional eigenspace $\mathcal{E}$. Since A commutes with $A'K$, A maps $\mathcal{E}$ into itself, and thus A must have an eigenvalue. Since A is nonscalar, the associated eigenspace $\mathcal{M}$ cannot be the entire space $\mathcal{X}$, and clearly $\mathcal{M}$ is invariant for $\mathcal{A}'$. This is a contradiction, and the proof is complete.

This theorem is indeed remarkable, and it leaves the invariant subspace problem in a peculiar situation. We shall return to this topic shortly, but first, we present a completely elemenatry argument, due essentially to M. Hilden, that establishes a special case of Theorem 3.

THEOREM 3′. *Every nonzero compact operator K in $\mathcal{L}(\mathcal{X})$ has a nontrivial hyperinvariant subspace.*

PROOF. If K is not quasinilpotent (i.e., if the spectrum of K is not the singleton $\{0\}$), then the eigenspace associated with any nonzero eigenvalue of K will be a nontrivial hyperinvariant subspace for K. Thus we may and do assume that K is quasinilpotent. We also suppose that K has no nontrivial hyperinvariant subspace, or, in other words, that the algebra $\mathcal{A} = \{T \in \mathcal{L}(\mathcal{X}): TK = KT\}$ is transitive. The proof now proceeds exactly as does the proof of Theorem 2 until the existence of operators $T_1, \cdots, T_n$ in $\mathcal{A}$ satisfying (1) has been established. Condition (1) tells us that to each y in $\mathcal{D}$ there corresponds at least one index $i = i(y)$, $1 \leqslant i \leqslant n$, such that $T_i y \in \mathcal{S}$. We apply this fact repeatedly as follows. Since $Kx_0 \in \mathcal{D}$, there exists an operator T_{i_1} in the set $\{T_1, \cdots, T_n\}$ such that $T_{i_1}Kx_0 \in \mathcal{S}$. Then $KT_{i_1}Kx_0 \in \mathcal{D}$, and there exists an operator T_{i_2} in the same set such that $T_{i_2}KT_{i_1}Kx_0 \in \mathcal{S}$. By induction, we obtain an infinite sequence $\{T_{i_k}\}_{k=1}^{\infty}$ of operators from the set $\{T_1, \cdots, T_n\}$ such that, for every positive integer k,

$$x_k = T_{i_k}KT_{i_{k-1}} \cdots KT_{i_1}Kx_0 \in \mathcal{S}.$$

Let ϵ be a positive number with the property that every vector x in $\mathcal{S}$ satisfies $\|x\| \geqslant \epsilon$, and let $M = \sup_{1 \leqslant j \leqslant n} \|T_j\|$. Then, since K commutes with each T_{i_k}, we have

$$\epsilon \leqslant \|x_k\| \leqslant \|T_{i_1} \cdots T_{i_k}\| \, \|K^k\| \, \|x_0\|$$

and

$$\epsilon^{1/k} \leqslant M\, \|K^k\|^{1/k} \|x_0\|^{1/k}$$

for every positive integer k. But this is a contradiction, since $\|K^k\|^{1/k} \longrightarrow 0$, and thus the proof is complete.

It is worth mentioning that, as of this writing, it appears that no one has succeeded in proving Theorem 3 without recourse to the Schauder-Tychonoff fixed point theorem.

We wish now to say a few words about the significance of Theorem 3 with respect to the invariant subspace problem, and for the purposes of this discussion, let us restrict our attention to the case that X is a separable, infinite-dimensional, complex Hilbert space H, since it is in that context that the most progress has been made on the invariant subspace problem. If $A, B \in L(H)$, let us write $A \longleftrightarrow B$ to mean that A commutes with B, and also write $A \sim B$ to mean that A is *quasi-similar* to B. (By definition, this means that there exist operators X and Y in $L(H)$ with trivial kernel and dense range such that $AX = XB$ and $YA = BY$.) Consider now the following diagram:

$$K \longleftrightarrow A(\neq\lambda) \sim B \longleftrightarrow C. \tag{2}$$

An interesting and obviously important question concerning this diagram is as follows. As K runs through the set of nonzero compact operators in $L(H)$, what subset of $L(H)$ does C run through? The point is this: if A is not a scalar and commutes with such a K, then, by Theorem 3, A has a nontrivial hyperinvariant subspace. Furthermore, by a theorem of Fillmore **[4]** and Hoover **[5]**, B also has a nontrivial hyperinvariant subspace, and thus C has a nontrivial invariant subspace. It is not known, as of this writing, whether the set of all C satisfying (2) for some appropritae $K, A,$ and B is equal to $L(H)$. In other words, it may turn out (although it does not seem terribly likely) that Theorem 3 *solves* the invariant subspace problem, at least for operators on Hilbert space! Thus there is no doubt that the diagram (2) is worthy of investigation. In the same vein, it is clear that every effort should be made to prove theorems similar to Theorem 3, and we now present two theorems of this nature. (Henceforth in this paper, X is once again an arbitrary infinite-dimensional, complex Banach space.)

THEOREM 4. *Let A be a nonzero quasinilpotent operator in $L(X)$, and suppose that there exists a bounded sequence $\{J_n\}_{n=0}^{\infty}$ of nonzero operators in $L(X)$ such that J_0 is compact and such that $AJ_n = J_{n+1}A$ for all n. Then A has a nontrivial hyperinvariant subspace.*

SKETCH OF PROOF. If A has only the trivial hyperinvariant subspaces, then the commutant algebra $\mathcal{A}'$ of A is transitive, and applying Theorem 2 to $\mathcal{A}'$ and the nonzero compact operator AJ_0, one obtains an operator A' in $\mathcal{A}'$ and a nonzero vector x in $\mathcal{X}$ such that $A'AJ_0x = x$. Thus, for every positive integer n, $(A'AJ_0)^n x = x$, and this leads to a contradiciton in a manner similar to that in the proof of Theorem 3′.

THEOREM 5. *Let K be a nonzero compact operator in $\mathcal{L}(\mathcal{X})$, and suppose that B is a nonscalar operator commuting with K. Suppose also that $B \in \{A\}''$ for some operator A in $\mathcal{L}(\mathcal{X})$, where $\{A\}''$ denotes the second commutant of A (i.e., $\{A\}'' = \mathcal{A}''$, where $\mathcal{A}$ is the algebra of all polynomials in A). Then A has a nontrivial hyperinvariant subspace.*

PROOF. The operator B has a nontrivial hyperinvariant subspace by Theorem 3, and since $B \in \{A\}''$, the commuting algebra of B is larger than the commuting algebra of A.

Even though this theorem is an easy consequence of Theorem 3, we include it here because it gives rise to the following interesting diagram:

$$K \longleftrightarrow B = f(A) \longleftrightarrow \{A\}' \longleftrightarrow A.$$

The point is that functions of A, if defined in any reasonable fashion, turn up in the second commutant of A. Thus, to conclude that A has a nontrivial hyperinvariant subspace, it suffices, in view of Theorem 5, to know that *some* (nonscalar) *function* of A commutes with a nonzero compact operator.

The next theorem to be established is stronger than Theorem 3. We have chosen this order of presentation because, unlike the proof of Theorem 3, the proof of the following theorem depends on a nontrivial fact from the theory of transitive algebras.

THEOREM 6. *Let $\mathcal{A}$ be a transitive subalgebra of $\mathcal{L}(\mathcal{X})$ and suppose that $\mathcal{A}$ contains a nonzero compact operator K. Then $\mathcal{A}$ is dense in $\mathcal{L}(\mathcal{X})$ in the strong operator topology.*

PROOF. Without loss of generality, we may suppose that $\mathcal{A}$ is closed in the strong operator topology, and therefore closed in the norm toplogy. By applying Theorem 2 to $\mathcal{A}$ and K we conclude that there exists a compact operator K_1 in $\mathcal{A}$ which has 1 as an eigenvalue. Since the spectrum of K_1 consists of at most countably many points with 0 as the only possible cluster point, there exists an open plane disc θ_1 containing 1 and containing no other points of the spectrum of K_1. If $f(z)$ is the analytic function defined to be identically 1 on θ_1 and identically 0 on a disjoint simply connected open set θ_2 containing the remainder of the spectrum of K_1, then $f(K_1)$, defined by the Riesz functional calculus (see [**3**, Chapter VII, §3]), is an idempotent operator whose range is the finite-dimensional eigenspace of K_1 associated with the eigenvalue 1. Furthermore, since $f(z)$ can be approximated subuniformly on $\theta_1 \cup \theta_2$ by polynomials $p(z)$ via Lavrentiev's theorem [**7**], and since A is uniformly closed, it follows that $f(K_1) \in \mathsf{A}$. Thus A contains a nonzero operator of finite rank, and it has been known for some time that this is sufficient to enable one to conclude that A is strongly dense in $\mathsf{L}(\mathsf{X})$. (This is explicitly stated in case X is a Hilbert space in Corollary 2 of [**9**]. Perhaps the best way to check its validity in the present context is to verify that Lemma 3.4 of [**10**] is valid in this broader context, and to use the fact that A contains a finite rank operator to apply Lemma 3.4 of [**10**].)

A result stronger than Theorem 6 has recently been obtained by Azoff [**2**] for operators on Hilbert space. Since its proof is somewhat technical, we refer the interested reader to [**2**] for more details.

The next proposition was pointed out to us by W. Arveson and R. G. Douglas.

THEOREM 7. *Suppose that* Y *is an infinite-dimensional, reflexive, complex Banach space with the property that every compact operator on* Y *is the norm limit of a sequence of operators of finite rank. Suppose also that* A *is a transitive, norm-closed, subalgebra of* $\mathsf{L}(\mathsf{Y})$ *containing a nonzero compact operator. Then* A *contains all compact operators in* $\mathsf{L}(\mathsf{Y})$.

PROOF. Just as in the proof of Theorem 6, we begin by applying Theorem 2 and the Riesz functional calculus to conclude that there exists a nonzero idempotent Q in A with finite-dimensional range Y_1. It follows easily that the algebra $Q\mathsf{A}Q$, regarded as an algebra acting on Y_1, is transitive. Thus, by Burnside's theorem [**6**, p. 276], $Q\mathsf{A}Q \mid \mathsf{Y}_1$ contains all linear transformations acting on the space Y_1, and, in particular, there exists an operator P in $Q\mathsf{A}Q$ such that $P \mid \mathsf{Y}_1$ is idempotent and has one-dimensional range. But then P itself must be an idempotent in A with one-dimensional range, and using the transitivity of A, the reflexivity of X, and the fact that A is norm-closed, it follows readily that A contains all operators on Y of rank one. Hence A contains all operators on Y of finite rank, and the result follows.

The following corollary might be thought of as a noncommutative Stone-Weierstrass theorem for compact operators.

COROLLARY 8. *If* K_1 *and* K_2 *are compact operators on an infinite-dimensional complex Hilbert space* H, *and* K_1 *and* K_2 *have no common invariant subspaces other than* (0) *and* H, *then the smallest norm-closed subalgebra of* $L(H)$ *containing* K_1 *and* K_2 *is the algebra of all compact operators.*

We close this exposition by remarking that although the above results in large part supercede the previous results on invariant subspaces pertaining to compact operators obtained by refining the Aronszajn-Smith technique, there is at least one theorem from **[11]** and **[1]** that does not seem to be covered.

THEOREM 9. *Suppose that* A *is an operator on a complex Hilbert space of dimension* $\aleph_0$, *and suppose that there exist a sequence of rational functions* $\{r_n(z)\}$ *with poles off the spectrum of* A *and a sequence* $\{K_n\}$ *of compact operators such that* $\|r_n(A) - K_n\| \longrightarrow 0$ *and such that* $\{K_n\}$ *converges weakly to some nonzero operator. Then* A *has a nontrivial invariant subspace that is also invariant under every rational function of* A.

BIBLIOGRAPHY

1. C. Apostol, C. Foiaş and D. Voiculescu, *Some results on non-quasitriangular operators.* IV, Rev. Roumaine Math. Pures Appl. **18** (1973), 487-514.

2. E. Azoff, *Compact operators in reductive algebras,* Canad. J. Math. (to appear).

3. N. Dunford and J. T. Schwartz, *Linear operators.* I: *General theory,* Pure and Appl. Math., vol. 7, Interscience, New York, 1958. MR **22** #8302.

4. P. Fillmore, *Notes on operator theory,* Van Nostrand Reinhold Math. Studies, no. 30, Van Nostrand Reinhold, New York, 1970. MR **41** #2414.

5. T. B. Hoover, *Hyperinvariant subspaces for n-normal operators,* Acta. Sci. Math. (Szeged) **32** (1971), 109-119. MR **46** #7930.

6. N. Jacobson, *Lectures in abstract algebra.* Vol II. *Linear algebra,* Van Nostrand, Princeton, N. J., 1953. MR **14**, 837.

7. M. Lavrentiev, *Sur les fonctions d'une variable complexe représentables par des séries de polynômes,* Actualités Sci. Indust., no. 441, Hermann, Paris, 1936.

8. V. Lomonosov, *On invariant subspaces of families of operators commuting with a completely continuous operator,* Funkcional. Anal. i Priložen **7** (1973), no. 3, 55-56. (Russian)

9. E. A. Nordgren, H. Radjavi and P. Rosenthal, *On density of transitive algebras,* Acta Sci. Math. (Szeged) **30** (1969), 175-179. MR **40** #6276.

10. C. Pearcy and R. G. Douglas, *Hyperinvariant subspaces and transitive algebras,* Michigan Math. J. **19** (1972), 1-12. MR **45** #4186.

11. C. Pearcy and N. Salinas, *An invariant subspace theorem,* Michigan Math. J. **20** (1973), 21-31.

12. P. Rosenthal and H. Radjavi, *Invariant subspaces,* Springer-Verlag, New York, 1973.

UNIVERSITY OF MICHIGAN

INDEX